DARK AGES

The looming destruction of the Australian power grid

Mark Lawson

DARK AGES

The looming destruction of the Australian power grid

Mark Lawson

Connor Court Publishing

CONNOR COURT PUBLISHING PTY LTD
PO Box 7257
Redland Bay QLD 4165
sales@connorcourt.com
www.connorcourtpublishing.com.au

ISBN: 9781922815347

Cover design by Maria Giordano

Cover image: Thomas Frye - Man Holding a Candle - B1977.14.11492 - Yale Center for British Art.jpg, Creative Commons Zero.

Printed in Australia

ACKNOWLEDGEMENTS

This book grew out of a suggestion by Rafe Champion a member of the Energy Realists of Australia that we co-operate on a book that would be a compilation of a series of briefing notes he was organising. Mainly intended for time-poor politicians, the notes debunk various green myths, particularly those concerning renewable energy.

Rafe eventually bowed out of the project which took on a different character, although aspects of the original structure remain, notably the points at the start of each chapter.

This book does not get into the details of pricing, attempts to forecast power generation and demand, mainly because most readers would quickly lose interest. Instead, I have tried to keep the discussion general and stick to the obvious points.

There is no discussion of the debate over gas prices and proposals to reserve gas production for the domestic market in this book, except to pour scorn on some of the policy ideas. The idea that Hydrogen is somehow the new LNG is treated with the contempt it deserves.

I have included a list various terms used in the book which I may not have fully explained on first reference, and certainly don't explain each time I use them.

TERMS USED

NEM – National Energy Market. This is the electricity market covering the East Coast of Australia, including South Australia and Tasmania. The grids for the Northern Territory and Western Australia are separate.

AEMO – Australian Energy Market Operator. This operates the National Energy Market as well as the wholesale electricity market for the WA grid.

Dispatchable – conventional power plants – that is, fossil fuel plants – can be turned off and on at will at the direction of a central dispatch, so they are dispatchable. Renewable power plants are not dispatchable. Other power plants on the network have to be adjusted to accommodate changes in the wind and sun.

GW – gigawatt. A whole lot of electricity. The whole NEM has somewhere North of 50 GW worth of generating capacity.

GWh – gigawatt hour. This is a whole lot of electricity storage. A battery capable of generating one GW (a very big battery) for one hour holds one GWh. Government ministers, officials, activists and journalists routinely confused GW with GWh.

Load following – power plants which can easily vary their output at direction are said to be load following. Brown coal plants cannot adjust their output easily. Gas turbines can be powered up and down to suit.

MW – 1,000 MW is a GW. 1,000 MWh is a GWh. Sometimes it is more convenient to use MW rather than GW as a measure of output.

Pumped hydro – a water battery. A pumped hydro project has an upper dam with hydroelectricity generators and a lower pool or dam. When power is required, water is let out of the dam through the hydroelectricity generators into the lower pool. Excess power is stored by pumping the water back up into the upper dam. Snowy 2.0 is, in effect, a giant water battery.

Australian attempts to convert to renewables basically have no parallel. To make matters worse, not only do policy makers want to shift to all renewables they want to throw away the fossil fuel back up power sources.

1

POWER MADNESS

For many years Australians have been told that the electricity grids that we all use every day, often all day, must transition to green power sources, notably the wind and the sun. Further, they have been told that a reliable power supply will be maintained through this transition and that electricity will be cheaper than the power delivered from the old network of coal and gas fired power stations.

To date the exact opposite has happened. Attempts to shift power production to this clean, green nirvana has done little more than increase power prices and reduce the reliability of the grid. A major, obvious examples of the trend occurred a few weeks after the election of a new Federal Labor government in May 2022. The electricity market covering the area was suspended for several days, wholesale prices sky-rocketed and major users (aluminium smelters and the like) had to be paid to stay off the grid to reduce demand. Consumers were asked not to use dishwashers. The resulting compensation bill for major users booted off the grid may reach $1.7 billion.

This is set to get worse. In 2022 the new Labor government estimated that power prices to increase by more than half, 56 per cent, over two years. That estimate has since been pegged back. However, as will be shown in this book, the grid for Eastern Australia is being so mismanaged that by the mid-2030s consumers may be glad of power at any price. The power stations on that grid have been aging for some time. Private investors have not built new, conventional (that is fossil fuel) plants for many years, and the government has built one. Now activists and state governments are colluding to throw away most of the still active fossil fuel power station – gas and coal power plants – some with decades of service life left in them, without bothering to put anything much in their place.

Wind farms and solar power plants are supposed to be built to do the job of coal and gas plants to be decommissioned, although this ignores extensive experience in both Australia and overseas that a network of renewable plants cannot deliver reliable power. Part of the problem is the newly recognised phenomena of wind droughts contributing to the bad habit of renewable power networks of going missing in action, often at crucial times. In any case, the renewable energy assets are not being built at anything like the rate required. Instead, at the time of writing, investment in the area has tanked.

This looming disaster could be averted by the application of common sense. The closure of some still quite serviceable coal plants could be delayed by several years, and a network of gas turbines could be constructed to back up the much vaunted network of solar plants and wind farms, that is, if and when enough of them are built to make a serious contribution.

But there is no indication that policy makers are even aware of the looming disaster. Instead, when interviewed by a media that also has shown little understanding of the problem, they talk about the installation of batteries and more photovoltaic panels on roofs; hydrogen plants, wind farms to be built off the coast of Victoria and almost any other nonsense going. This is about as useful as talking about unicorns and pots of gold at the end of the rainbows, but never seems to be challenged.

Nor does anyone seem to understand that Australia's East Coast grid is being dragged into basically unknown territory – making a developed nation's grid all renewable with just wind and solar power plants, plus a little hydro power. Norway has an all renewable network but that is due to the country's vast hydro power assets. Hydro power counts as renewable while not being intermittent. In fact, as an energy source it is higher value than coal, gas and even nuclear. If we could all switch over to hydro power there would be no problem, but we cannot. New Zealand also has a high renewable penetration but, again, that's mostly from hydro with a healthy contribution from geothermal (heat from volcanically active areas used to drive turbines), another renewable that is high value in that it can be turned on and off at command.

No, Australia is a mainly coal and gas powered grid which policy makers are trying to convert to renewables and, to make matters worse, it is isolated and thinly spread grid. Europe has a patchwork of interconnected grids, most of which have many times the generating capacity of the Australia networks, all packed into an area roughly equivalent to the Australian Eastern grid. This makes for more opportunities to share power, although even with that level of interconnection and grid density there have been major problems. Australian attempts to convert to renewables basically have no parallel. To make matters worse, not only do policy makers want to shift to all renewables they want to throw away the fossil fuel back up power sources.

This has not happened anywhere and is madness.

Underlying all this is the endless, relentless demonisation of fossil fuels, including active campaigning against new gas wells and coal mines and the construction of any new plants using those dreaded fuels, all in the name of climate. This means that the Australia's once reliable fleet of coal-fired power stations is now starting to age – they are going off line more often and proving harder to maintain, which leads to all the problems noted earlier. As noted, renewable power sources are meant to pick up the slack but have major limitations and are not being built fast enough.

A far from surprising result of these trends has been a decline in the reliability of grids and an increase in power prices, as there is far less supply in the system. At the same time coal and gas prices have increased – a trend greatly exacerbated by the Ukraine-Russian war which has curtailed gas supplies from Russia – and those higher prices have further boosted power bills.

To make matters worse, the urge to control emissions which has brought all this on is already a long-lost cause. No matter what activists may say, efforts at controlling emissions world wide have proved an abject failure – a point discussed in more detail in this book. Even a slowing in the rate of increase of emissions growth may not occur, but then a lot depends on China. Otherwise, and in effect, all the money to be spent, all that effort and pain to be inflicted on Australian power consumers in the name of climate will be for nothing.

Australia is not alone in putting itself on the rack for no reason apart, perhaps, from making activists feel better about themselves. Power is also in desperately short supply in Europe with consumers in Germany trying to avoid crippling bills for gas and electricity by burning wood, Polish citizens queuing for hours to buy coal and businesses in the UK facing closure because they cannot pay their power bills. But then what can anyone say when the powerful German green movement forces its government into taking stupid decisions such as closing all its emissions-free nuclear reactors, following the disaster that overtook the reactor at Fukushima in Japan. The Fukushima disaster was the result of an enormous tidal wave and Germany is not subject to tidal waves, but green activism proved so powerful that the nuclear plants were closed anyway.

However, very little about the energy debate makes any sense.

During the May 2022 Australian Federal election, for example, all political parties abandoned the eccentricity and wishful thinking over renewable energy that had permeated political debate up to that point and descended into madness.

All sides talked off achieving net zero emissions by some future date when it is already clear that the goal is an unachievable fantasy – after decades of effort renewable sources account for just a few per cent of the world's total energy consumption (including fuel for transport, gas for heating and so on). The winning Labor party spoke of ensuring that 89 per cent of all new car sales would be electric and 15 per cent of all cars on the road would be electric by 2030, although both figures are clearly absurd. Perhaps 3 per cent or so of new car sales are EVs, and the target date is just seven years away.

Even if significant sales of electric cars could be achieved the result would be a considerable strain on electricity grids already struggling with the problems of an aging fleet of coal and gas power plants with no new replacements in sight.

It gets worse.

During his winning speech on election day, Prime Minister Anthony Albanese talked about making Australia a green "energy superpower" when academics have already pointed to huge problems with that concept. Australia is an energy superpower in coal and LNG, as it has reserves of both gas and coal which can be dug up and sent overseas using long-accepted technology. Not so with renewable energy as that can be generated anywhere using cheap Chinese equipment. Australia has few advantages in the area and is a long way from any potential customers, with no viable means of transporting the power over the huge distances involved at anything like a reasonable cost.

Hydrogen is certainly not an answer to the problem of energy transport. In fact, it is difficult to heap enough scorn on proposals for its use. Under the most optimistic projections hydrogen will not be a part of the energy scene for decades, if ever. Yet the election campaign featured promises of so many millions to be spent on the likes of hydrogen refuelling stations – a straight waste of taxpayer's money.

But then, as noted earlier, the whole emissions debate involves spending billions and imposing more billions in costs on the economy for no gain of any kind. Even if we assume that Australia's emissions reduction effort is matched by other countries, which plainly isn't happening, it is all but impossible to construct a cost benefit analysis to justify the billions being spent. Yet almost all commentators seem to have lost sight of that basic point, in favour of cheering on all emissions reduction efforts irrespective of the costs. After all, their jobs remain secure.

Trying to counter the evident, utter nonsense prevailing in this area is akin to shouting into a cyclone, but this book may help a little. We can only try and hope that someone, somewhere, is listening.

2

VANISHING CAPACITY

* The brown coal era was one of cheap power.

* That era has come to an end with the general shift towards wind and solar power and away from coal plants.

* Wind and solar plants cannot be controlled in the same way as conventional (that is, fossil fuel) plants can be.

* Australia's network of reliable coal fired plants are all aging and no plants are being built to replace them. Now the bulk of them are to be forced out of service by the middle of next decade.

In the grand old days – perhaps thirty years ago – each state had its own power supply authority and its own network of power plants, including huge brown coal plants such as the Loy Yang plants in the Victorian region of Yallourn, built next to major deposit of brown coal (often known as Lignite).

As brown coal is typically not exported and reasonably easy to mine it is cheap to local users. This cheap coal was turned into cheap power in enormous coal-fired generators. But the big brown coal plants were never the whole story of electricity supply, because their output cannot be varied very much. Once at a certain output level they are left that way. If they have to be shut down for any reason, they can take a whole day to restart.

But demand for power can vary quite a bit during the day. There are distinct peaks in the morning, when families have breakfast and start the day, and in the evening when everyone gets home and turns on stoves, heaters, washing machines and the like. Most of this is predictable at least to a certain extent, although there can be surges. During very hot days in Australia everyone will turn on their air conditioners and demand goes through the roof. During the FA cup in England, so the story goes, demand spikes at half time. Why? Because families across the country have been watching the game on television and at half time everyone decides to put on the kettle for a cup of tea.

Grid operators have to balance supply and demand as near as possible 24 hours a day, seven days a week, so that households across Australia get a more or less steady 240 volts of alternating current at the power plug. If there is too much supply and not enough demand the voltage increases, parts of the system overload and, hopefully, circuit breakers will trip rather than have vital sections burn or blow up. High voltages will also greatly upset the millions of domestic appliances plugged into the grid.

The opposite condition of too much demand and not enough supply will result in parts of the grids not getting enough supply (brown outs) or none at all (black outs). Consumers will not be able to cook meals, wash clothes, watch television or, perhaps most serious of all these days, not be able to get onto the internet.

Another problem is that of maintaining current frequency. As noted, the standard household supply is an alternating current as opposed to direct current which always flows in one direction (that is the positive side remains positive and negative side negative). In contrast, alternating current continually reverses direction. In Australia the standard current reverses sixty times a second, so it has a frequency of sixty hertz. We will not discuss balancing frequency further, but it is worth mentioning here as one of those nasty details that only electrical engineers really care about, until activists start insisting that the network has to be supplied by intermittent energy sources.

Back to brown coal plants, and the related black coal plants. Australia also has lots of black coal. As the output of these cannot be varied much other types of generators are used to fill the gap between the coal plant output and varying demand. We will only mention two here, open cycle gas turbines and hydropower. Gas turbines are something like very large jet engines. They can be powered up and down very quickly. As for hydropower, everyone would be familiar with the concept of water from large dams being channelled through an electrical turbine to create electricity.

Because gas and hydro generators can be powered up to full output quickly (very quickly in the case of hydropower) they are termed peaking plants. They are switched on to accommodate peaks in the power supply and powered down again when the peak passes. As these plants can be switched on at the direction of a central dispatch, they are also called "dispatchable" and, as they can also be set to most outputs in-between full power and no power at all, they are said to have good "load following" capabilities. They can be quickly adjusted to suit both changes in demand and supply so that the overall grid voltage remains stable (within a few per cent).

There is another complication. Grid operators have to be ready to patch over any sudden problem in the grid, such as an accident at one of the power plants or sudden surges in demand. Typically, a grid will have a certain amount of power in reserve, including a spinning reserve ready to be hooked up at a moment's notice plus additional amounts in different states of readiness. The amount in reserve required at any point is calculated by arcane rules, although one rough rule of thumb used to be that the capacity in reserve would be equal to the largest generating unit on the grid.

Wind and solar plants cannot be directed by any central grid dispatch or control and cannot be used for reserving. When a lot of wind power or photovoltaic power comes onto the grid the collective output of the conventional power plants has to be adjusted to suit the influx. Individual intermittent generators may be directed to stay off the grid for a time in order to control the increase, but otherwise the conventional generators have to be adjusted, somehow, to the new supply.

If a lot of wind and sun powered generators suddenly stop working, because its night or the wind has stopped blowing, then conventional plants have to be powered up to fill the gap. As discussed later in the book, no conventional power plants may then mean no electricity at certain times.

The huge number of rooftop photovoltaic panels throughout Australia illustrate the problem. There are now perhaps three million individual rooftop installations in Australia thanks to years of government incentives, but they typically do not obey the orders of a central dispatch. In the middle of a mild but sunny day with the rooftop panels producing at full rated capacity but not much demand, grid operators may have real problems shutting down enough of the conventional generating capacity to accommodate the additional supply. When the evening peak hits, all that conventional generating capacity has to be restarted to keep the grid going.

To deal with the problem over supply in February 2022 WA followed the lead of South Australia in bringing in regulations requiring all new solar panel installations to have inverters that will switch off the panel at the direction of the grid authorities. (Victoria now has a similar rule for new installations.) This would seem to be too little, too late considering that the problem had been known and widely publicised for many years before the state governments took action. However, it is better than nothing.

This is the heart of the problem. State and Federal governments have insisted on encouraging the development of renewable energy which can cause major problems in managing the grid, while at the same time forcing still functioning coal and gas powered plants out of service, with no replacements in sight. They are sowing the seeds of a major disaster – worse, they are doing so while denying that there will be a problem.

In September 2022 for example, billionaire activist Mike Cannon-Brookes and the Queensland government between them managed to engineer the early closure of a huge chunk of the reliable coal-fired power supply of the Eastern half of the continent over the next 13 years.

The situation was bad enough before Cannon-Brookes and the state governments of Queensland and Victoria became involved. As has long been known, the 1.8-gigawatt Liddle power station in NSW run by energy company AGL will close in 2023, and its 2.6-gigawatt Bayswater plant will cease operations between 2030 and 2033. In addition, another major energy company Origin Energy will shut its 2.8-gigawatt Eraring coal-fired power station in 2025 (this is being reconsidered at the time of writing), and Victoria's Yallourn power station (1.48 GW, brown coal) is scheduled to close in 2028.

Those closures are now to be joined by many others.

After a sustained campaign by Cannon-Brookes, who became AGL's largest shareholder with the express purpose of getting the energy giant to accelerate closure of its coal plants, AGL has also announced that will shut the shut the 2.2-gigawatt Loy Yang A power station in Victoria's La Trobe Valley in 2035, a decade earlier than planned.

At about the same time as the AGL announcement, Queensland Premier Annastacia Palaszczuk declared that her government would end the use of coal power in the state by 2035. Her government owns eight coal fired plants in the state (the power plants in NSW and Victoria are privately owned), the newest of which is the 30 year old 1.4-gigawatt Tarong station, which will now close more than a decade ahead of schedule (other, privately held plants will remain open). Victorian premier Dan Andrews hopped onto this bad wagon in October by declared that his government will introduce tough new emission targets that are likely to end coal powered electricity generation in the state by 2035. The one remaining coal plant to be affected is Loy Yang B which generates a little more than 1 GW.

To replace this gaping hole in generating capacity, both state governments announced a host of renewable energy and storage projects. The Palaszczuk government intends to develop a $62 billion renewable energy "super grid" which includes a new transmission line and two new pumped hydro projects. The Victorian government has declared it will revive the old State Electricity Commission which built the original state grid, but this time as a renewable energy agency with $1 billion to develop 4.5 GW worth of renewable energy projects. It has also announced an increase its renewable energy storage target to – and this is a direct quote from the press release – "6.3 GW by 2035".

These initiatives will be discussed in more detail elsewhere in the book but the Victorian announcement about storage is hardly reassuring as GW is a unit of output. The unit of storage is GWh. A generator rated at 1 GW which produces that one gigawatt for 10 hours has produced 10 gigawatt hours. Politicians, commentators, activists and government press releases routinely confuse GW and GWh. The Queensland government announcement at least contains more details than the Victorian announcement but makes exactly the same mistake in confusing gigawatts and gigawatt hours.

Not to be left behind by its Labor counterparts, the NSW government has a system of five renewable energy zones beginning with a pilot in the central west of the state. The government has tipped in $380 million and is hoping for more than $20 billion worth of private investment.

Whatever the result of all these announcements the problem

remains that a major part of the country's reliable capacity is about to vanish with no replacement in sight. No other coal or gas plants are being built apart from one peaking gas generator which the new Labor government is doing its best to handicap, as discussed later in this book. The state governments seem to be relying on private development to create the truly vast amount of renewable energy and storage that will be required to make up for the generating capacity to exit the market. The trouble is, as we shall see, private investors are proving slow to respond, with various players in the market noting that what the grid really needs is more reliable supply, rather than simply more renewables.

In September 2022, before the AGL and Queensland announcements, the head of the Australian Energy Market Operator which runs the east coast grid, Daniel Westerman, declared to the Australian Financial Review that "events of the winter", meaning a major crisis in June, "have reinforced the need for Australia to continue to invest in the transition towards firmed renewables". By firmed renewables he meant pumped hydro and perhaps batteries, but reports produced by the AEMO have mentioned the dreaded word "gas".

Details of this crisis and the ins and outs of market interventions and the problems with price caps will not be examined in this book. We need only note that the crisis involved a sudden increase in demand due to a cold snap and more of the coal generators being offline than expected, thanks to the general aging of the power plant fleet, low supply from wind and solar power generating assets, and sky rocketing coal and gas prices.

All the players involved in this saga are likely to be better prepared next time around, but the major problem remains that more of Australia's reliable coal and gas fired plants are going out of service and governments, for ideological reasons, are refusing to consider building power plants that might replace them or to create the conditions by which private investors may build plants which produce dispatchable power.

In the meantime, the power grid with its collection of aging fossil fuel plants will stagger along somehow, at least until the Liddle plant ceases operation in 2023. As previously noted, grid operators can use a technique called "demand management" which involves paying major users such as aluminium smelters to stay off the grid during a crisis. But this still leaves the basic problem, dealt with later in this book, that when wind goes down it may do so over a wide area and may be down for days. Just what the Federal and state governments intend doing during what are now being called "wind droughts" is not being discussed.

Instead of acknowledging this looming problem, commentators descend into fantasy about how more renewables and extensive use of hydrogen will fix the problem. It seems that consumers must wait until they are left in the dark in freezing homes for extended periods before policy makers finally concede that renewables might not be the answer to everything.

Addenda

An object lesson in how not to manage a grid has been provided by South Africa. There the government has allowed the network of coal power plants that provide most of the nation's electricity to age badly, just as the country's households were connecting to the power supply in increasing numbers, while not building even green power plants as replacements. According to the International Energy Agency in 2020 about 7 per cent of the nation's power came from renewable sources.

The South African government seems to have known about the looming crisis for some time, without taking the strong action required. South Africa's state-owned energy company, Eskom, is reported to be both heavily indebted and strike prone.

3

WIND AND OTHER CATASTROPHES

* Wind droughts, that is periods of flat calm, will prove an increasing problem as more of Australia's power comes from wind generators.

* These droughts are caused by high pressure systems that move across the continent and can cover most of the Eastern Australia. There is as yet no way to forecast the windless periods resulting from those systems and storing enough power to keep grids operating during such periods is difficult and costly.

* With coal plants falling out of service, the increased reliance on wind and sun may result in large sections of Australian grids being left without power for days.

They are called a wind droughts. The equivalent increasingly used phrase in German is dunkelflaute, the literal meaning of which is dark doldrums or dark lull – periods when both the sun does not shine and the wind does not blow. In the days of sailing ships sailors were used to dealing with both storms and periods of calm weather which could last for days, if not weeks, but no one concerned themselves with the size of the area that remained windless. Now the evidence is that when wind dies, it does so over a very large area.

To illustrate this problem, we can look to Europe and the UK where there are plenty of wind farms spread over an area similar in size to that of the Eastern half of Australia. Late in 2021 as delegates in the annual climate summit, held in Glasgow that year, were noisily demanding more renewable energy, the UK had to turn on mothballed coal power plants because of one such wind drought.

In an article on the Australian edition of the academic site The Conversation published in October 2021 a researcher in climate risk analytics at the University of Bristol in the UK, Hannah Bloomfield, said that the period of still weather badly affected wind generation across the continent. She said that the renewable assets of the UK-based power company SSE produced 32 per cent less power than expected.

Dr Bloomfield also commented that these "wind droughts" can be classified as an extreme weather event, like floods and hurricanes. Stagnant high atmospheric pressure systems over central Europe, lead to prolonged low wind conditions and those conditions may be "difficult" for power systems in future.

(High and low pressure systems, large areas where the air pressure is above or below standard, are a feature of the atmosphere. People will mostly not notice these changes but a barometer, which measures air pressure, can be a pointer to changes in atmospheric conditions which they will notice. If pressure starts falling a storm may be on the way. High pressure systems are associated with calmer weather.)

Dr Bloomfield tried to put a positive spin on the matter in the article by stating that it was important to understand just how such events occur, so that they could be forecast and the grids prepared for them. But even if forecasting was possible, and researchers are still trying to do this reliably with long recognised extreme events such as floods and cyclones, it is difficult to see just what can be done about wind droughts beyond turning the fossil fuel generators back on, especially as there is evidence that such droughts can continue for weeks in Europe.

In 2018 a wind drought in the UK meant that wind made no

contribution to the UK grid at all for nine days, according to media reports, and only slight contributions for another two weeks. In the wind drought of late 2021 noted earlier, there were days when wind made no contribution at all and it affected the whole continent, sending power prices through the roof.

Energy realists also note that it costs around £5 million ($A8.8 million) just to switch on a single coal fired plant for the duration of a wind drought and, when there is too much wind in the UK, it costs even more to pay wind farms to shut down so as not to overload the grid. These balancing costs are estimated to have amounted to £1.8 billion pounds for the UK in 2020/21, with every indication that the problem is getting worse not better.

An examination of the European wind records by the Dutch academic authors of *A Brief Climatology of Dunkelflaute Events* (in the journal Energie, 2021, available online), showed that almost all such lengthy events (more than one day) occurred in November, December, and January for European countries. This is during the European winter when power is needed for people to keep warm.

"On average, there are 50–100 h of such events happening in each of these three months per year," the paper said. A glimmer of hope for activists was that the paper also said that although the probability of the same event occurring in two adjacent countries was high, it was much less probable for the same event to occur in all the countries surrounding the North and Baltic seas. It doesn't happen everywhere all at once. More interconnections between the different national economies may at least reduce the size of the problem.

In another paper, *Geophysical constraints on the reliability of solar and wind power worldwide* (Nature Communications, 2021) a team lead by Chinese scientist Dan Tang reported on the modelling of a range of different contributions from solar and wind sources to the energy mix for different countries. It concluded that "assuming perfect transmission and annual generation equal to annual demand, but no energy storage, we find the most reliable renewable electricity systems are wind-heavy and satisfy countries' electricity demand in 72–91% of hours."

This was improved to 83–94 per cent by adding 12 hours of storage, or about the amount that is in various stages of development on Australia's east coast. However, the paper also notes that "even in systems which meets more than 90% of demand, hundreds of hours of unmet demand may occur annually". In other words, power users will still be left in the dark for long periods.

Like all modelling work the conclusion depends on a number of simplifications such as fresh water always being available for pumped hydro and that the dams will be fully recharged between each wind drought. These points will be discussed in more detail in other parts of this book.

Australia may well be more fortunate than Europe as our wind droughts do not seems to last nearly as long as those in Europe – as far as anyone knows – so we will not need as much very expensive power storage. But we will still need a lot.

Despite this being an obvious problem with renewable systems which this country is doing its best to adopt, Australian academics have shown little interest in the issue of long periods of calm weather and overcast days. Studies of the problem that have been done in the past year or so do not help as they assume far larger renewable energy networks than are being built, or that far more turbines than necessary for routine operations will be built throughout the whole network, resulting in more power production during wind droughts. In any case, the resulting claims that the entire east coast grid will be able to get by with storage of just a few hours worth of power are hard to take seriously.

Somewhat more convincing as a starting point in assessing at least the size of the problem is the work of a group of concerned citizens some of whom have coalesced around a group called the Energy Realists of Australia. These include Sydney residents Rafe Champion, founder and CEO of non profit energy and educational consultancy Energy Agnostix, Peter Feros and John Morgan. Anton Lang, blogging as Tony from Oz, has amassed a series of wind statistics starting around 2008, and has posted thousands of reports on power generation in SE Australia.

These concerned citizens have analysed the figures for wind production on the National Energy Market, which is the grid for the Eastern half of Australia, using statistics readily available from the grid manager, the Australian Energy Market Operator.

An independent analyst working with the Energy Realists who did not want to be named, has produced a spreadsheet of wind droughts for each year from 2011 to 2020. A feature of his analysis is that from 2015 through to 2019 the wind droughts are quite short, the longest being about three hours at most, with capacity factors (average output) for wind farms in the NEM of perhaps 7-10 per cent. But all other years have serious wind droughts, the longest being 74 hours in 2011, plus three more of longer than three days in the same year, all with wind farms showing an average capacity factor of 3-6 per cent. For 2020, the analysis found the longest wind drought for the year was 33 hours with plenty more of at least nine hours.

These figures indicate that a system using solar and wind power might hang together somehow, or at least not be noticeably inadequate, for a few years (long enough for those involved to collect their environmental awards) and then fail to produce more than a fraction of the power required for days at a time for years after that. A system relying on intermittent energy is not only volatile during the year, but its performance would seem to vary from year to year.

A group which runs the web site *Stop These Things*, posted an analysis on its web sites of the AEMO figures for June 2020, a winter month, which is worth quoting as an indication of the vast problems faced by those who have to manage wind production. The site notes that there are dozens of occasions when the entire wind fleet battled to deliver more than a tiny fraction of its combined capacity.

"Spread from Far North Queensland, across the ranges of NSW, all over Victoria, Northern Tasmania and across South Australia it routinely delivers just a trickle of its combined notional capacity of 7,728MW (7.728 GW).

"Collapses of over 3,000 MW or more that occur over the space of a couple of hours are routine, as are rapid surges of equal magnitude, which make the grid manager's life a living hell, and provide the perfect set up for power market price gouging by the owners of conventional generators, who cash in on the chaos," the post says.

Detailed work has also been done by electrical engineer Paul Miskelly, who wrote a paper *Wind Farms in Eastern Australia – Recent Lessons* (Energy and the Environment 2012, available online). The paper analysed the output of 21 wind farms connected to the Eastern grid spread over a very wide area, in fact the "most widely dispersed, single interconnected grid" at that time, and found that there was no apparent reduction in the volatility due to the spreading of wind farms. There was no smoothing of output. Instead, the collective output of those wind farms can vary from zero or near zero to substantial output in little more than an hour.

Contacted as part of the research for this book Miskelly said that the addition of more wind generations in the area since 2012 had done nothing to improve smoothing. He points to Monthly Wind Power Graphs on a site Aneroid Energy https://anero.id which compiles wind production statistics from the AEMO. The graphs showing total wind output are virtually the same as graphs in the 2012 paper.

"The result is entirely in line with my prediction, resulting from observation that, quite frequently there occur large, widespread high-pressure systems that result in little wind anywhere right across the Eastern Australian grid," Miskelly said.

When the Aneroid Energy site was inspected for a time during the final stages of writing this book, the wind generator output for all of the NEM varied from 2.2 GWs down to 295 MW (0.295 GW), or just 2.5 per cent of the total generating capacity of the wind farms. A handy and comprehensible real time meteorological chart also on the site shows that the production low coincided with a high pressure system over South Australia.

These windless high pressure systems typically drift across Australia from East to West and may linger in the continent's South East corner, where the bulk of the wind generators are, for hours or days, causing major wind droughts. As noted above major wind droughts can be rare but they do happen.

In fact, as I have noticed looking at the wind output data on Aneroid and other sites over the years wind production in South Eastern Australia is characterised by an irregular saw-tooth pattern. The troughs in this pattern can pass relatively quickly but sometimes they don't.

A typical response to this obvious problem by activists is to point out that building more turbines in all areas will lift the base so that 0.295 GW becomes 29.5 GW. Surely, the effects of the high pressure system will not extend all the way to Queensland, so why not build some more wind farms there? Also, solar generation can always be relied on to produce something even during cloudy days and is far more predictable, so the argument goes, with a distinct peak during the middle of the day. Build some more solar generators and just a few hours of storage should then smooth out all these peaks and troughs.

Readers will note that during a wind drought, when the sun is shining, grid operators will have to deal with a power glut during the day and a major deficiency at night. Spot checks of the Aneroid site does not show much difference between the average output of wind farms in Queensland, as opposed to those in Victoria and, in any case, the over building argument means that Australia needs not only enough wind farms and solar power plants to meet immediate demand – capacity that is not being built in anything like the numbers required – it will also need a whole lot more to cover major changes in wind patterns, plus storing enough power to keep the grid functioning through nights.

Another major problem is that sunshine can also come and go, so where are the detailed analyses of both wind and solar production from existing meteorological statistics? Activists tend to wave their hands and say that it will all work out – the god of renewables will provide – but it would be nice to have detailed analysis from independent parties to back up those reassurances.

Apart from the vast additional expense required to generate supposedly cheap, green electricity, the approach of building lots more wind farms onshore has distinct problems in the form of planning regulations. In Victoria wind turbines, by and large, cannot be built within one kilometre of an existing house, or in environmentally sensitive areas such as national parks and are excluded from large areas in central Victoria, mostly around Melbourne. That still leaves a lot of area, especially out in the plains of Western and North Western Victoria, but we will need a lot of wind farms. Victoria wants to build offshore wind farms, which are known to be more expensive than on shore installations, in large numbers, but there is no real indication that this will help very much with the problem of wind droughts.

In any case, to generate power the wind farms and green energy projects have to be built in vast numbers, that means we need major increases in the number of such projects in the development pipeline. The opposite has happened. Figures on renewable energy projects complied by the Clean Energy Council shows that just 17 were completed and commissioned in 2022, representing 1,248 Megawatts (MW) of installed capacity, as opposed to 48 completed in covid-stricken 2021 adding up to 4,589 MW, and 3,205 MW worth of projects in 2020. These figures are even less impressive when it is remembered that wind projects typically have an average output of about one third of stated capacity. The average output for photovoltaics is somewhat less, but it is better for projects using the likes of solar concentrators.

There is no indication investment is improving. Wind farms under construction listed by the Victorian Department of Transport and Planning in early 2023 amount to just 864 MW in installed capacity – an effective average output of perhaps a paltry 300 MW or so.

A part of the reason for investment in this area falling off a cliff, despite all the talk, is that markets did not do well generally in 2022. A count of Initial Public Offerings on the securities exchange by professional services firm HLB Mann Judd shows that the number of new IPOs fell by 48 per cent in 2022, and total funds raised collapsed 91 per cent.

Another perspective is provided by lobby group WindEurope, which in January declared that orders for new wind turbines in Europe fell by 47 per cent, or nearly half in 2022 compared with the previous year. WindEurope complained about government interference in the European markets, but also noted that "inflation in commodity prices and other input costs has raised the price of wind turbines, by up to 40 per cent over the last two years". Revenue had not kept pace with costs.

The Green Energy Council blamed this poor Australian result on the Federal Liberal-National government. But that government went out of power in May 2022, and the green-mad Labor governments of Victoria and Queensland have been in government for years, with the Liberal regime in NSW also proved distinctly green friendly.

In other words, private investors are not proving as enthusiastic about green energy as hoped and, with coal power plants falling out of service, there is a good chance that consumers may be left without power of any kind. This should be ringing alarm bells and at the very least should result in a major rescheduling of closures, but the figures have been shrugged off. The gods of climate will provide.

Let us return to the issue of trying to keep the grid stable, assuming for a moment that it will have enough green power to meet demand. For the sake of argument and to give us an idea of the size of the storage problem, let us assume that we will need to store power for a day and a half to tide the grid over these inconvenient windless periods. The NEM has North of 50,000 MWs (50 GW) of generating capacity. If we also assume that an average of half that is used (more during peaks and less during troughs) during any given period, then the market will need around 900,000 megawatt hours (900 GWh) to get through a 36 hour drought. Note that this figure is just to give an idea of the size of the problem, rather than a formal estimate.

An Integrated System plan produced annually by the Australian Energy Market Operator, looks at the likely future requirements of a grid with heaps of renewable energy. The 2022 ISP estimates that by 2050 the Eastern grid will require 640 GWh of dispatchable storage in all its forms, including pumped hydro and batteries. But this future grid will also have 10 GW of gas powered generation for firming (reliable) generation and peak loads. However, the activists now in control in Victoria and Queensland have completely ruled out gas. Without that gas generation vastly more storage will be required.

Those guesstimates show that batteries will be of little use. The Hornsdale Power Reserve battery built in South Australia in

2017 with considerable fanfare, for example, cost $90 million but stores just 125 MWh. As discussed in another chapter Hydrogen storage is even less efficient and far more difficult to manage than batteries.

Dams or pumped hydro projects which uses vast amounts of water in dam as a store of energy are better, if expensive (there is also more discussion of pumped hydro in another chapter). The pre-eminent such project is Snowy Mountain 2.0, which is turning into a ruinously expensive white elephant. The original cost estimate was $2 billion but the project is now expected to cost three times that and still counting. When this money hole is finished, however, and assuming it can find enough fresh water to fill its various reservoirs it should store about 350 GWh. Three such projects might then tide the NEM over a 36 hour period, although only one is being built, plus the equivalent of perhaps another in various pumped hydro and battery projects promised by the state governments. (Figures can be difficult to establish.)

As noted in the previous chapter, the Palaszczuk government in Queensland has announced that it will develop a $62 billion renewable energy "super grid" which includes a new transmission line and two new pumped hydro projects. About half of that investment is expected to be public money, including $9 billion from the state government and (hopefully) the rest from the Federal government.

Online searches show that one of the Queensland projects, which uses the existing Borumba dam near Gympie in south-east Queensland is expected to store 48 GWh. The second, the Pioneer-Burdekin project near Mackay, is larger than Borumba although no storage figure is available. The two together have been billed as being larger than Snowy 2.0. At the time of writing both projects were still at feasibility stage, meaning that there is a long way to go, particularly for Pioneer-Burdekin where there is no existing dam.

The Victorian government announcements indicate that its projects have even further to go. Three battery projects in various stages of development amounting to about 1 GWh have been mentioned, and the state government is tipping in $167 million of taxpayer money, although there are no details about what the money is to be spent on. Otherwise, the announcement seems to be a statement of intentions about getting the private sector involved in building stuff.

In October 2022, the Victorian government also announced that it would revive the old State Electricity Commission – sold off by a previous state government many years ago – but this time as a renewable energy agency with $1 billion to develop 4.5 GW worth of renewable energy projects. Again, these are fine words but even $1 billion does not seem like very much to replace many billions of dollars' worth of coal fired plants going out of commission. Media reports suggest that the Victorian government is hoping to get private investors interested in renewable energy projects, notably the massive superannuation funds. That might make up some of the shortfall if the super funds can be persuaded to invest their member's money in projects where the governments involved get their electric measurement units mixed up.

There is no shortage of suggestions about how this or that existing dam close to established power lines can be quickly adapted into a pumped hydro system simply by constructing a lower reservoir. The strong possibility that even for these simplified projects just the environmental assessments and feasibility studies alone may take years to complete; that power requirements would have to take a back seat to the dam's actual purpose (irrigation), especially during water droughts, and that the power lines may have to be upgraded at considerable expense are all waved away, often with a show of irritation.

In the meantime, while governments make muddled announcements about what they may be going to do, it seems a fair bet that the eastern grid will be short of a lot of generating capacity when all the planned closures of coal plants occur, and even shorter of storage capacity.

For the figure of the equivalent of three Snowy Hydros mentioned above is just a bare minimum. Grids have to be organised to handle worse case scenarios. What happens if a far longer wind drought occurs? What happens if one wind drought is followed by another before the dams have been able to completely recharge, so to speak, by having the water pumped back into them? What happens when there is a water drought and so no water to fill these pumped hydro facilities? (As this is being written. Norway is unable to supply power as it would like to other countries, due to a bad drought affecting water level in its dams.)

Of a little more use in operating a grid in these strange times is the concept known as demand management. This means paying major users such as aluminium smelters to stay off the grid when generation fails. Would the storage planned to date and demand management be enough?

Hold on, what about worse case scenarios such as a burning hot day, when everyone turns on their air conditioning, which also happens to be a calm cloudy day? What about a lengthy, searing, region wide heatwave with cloudy days and hot nights, when wind is generally low without the region necessarily being in a wind drought? Building enough storage to take account of all such scenarios is impossible. Some fossil fuel backup is still required.

Before being thrown out of office in May 2022 the Morrison government at least made some effort to replace the expect substantial loss of firm capacity by commencing a $600 million 600 MW gas fired plant in NSW's Hunter region. Initiated after it was announced that the Liddell station would close and to be operated by the Snowy Hydro Authority, the project was pushed through in the teeth of opposition from government and activists.

The Labor government elected in May 2022 reversed its opposition to the plant but only after insisting on conditions which will hobble its effectiveness. For the government had decreed that 30 per cent of the gas used by the generator must be green hydrogen from day one of operation (the planet should be operating by the time this book is released). Further, all of the plant's gas supply has to be hydrogen by 2030, or in just eight years time.

The plant can run on a mixture of natural gas and hydrogen without modification, as opposed to the vastly more convenient straight natural gas, but would have to be substantially rebuilt to run solely on green hydrogen. The problem is that there are no sources of green hydrogen in that region or anywhere else in Australia. During the election campaign, Labor declared that it would set aside another $700 million for the Snowy Hydro Authority to make green hydrogen on the site. In other words the government wants to build a renewable energy power plant on site to use scarce fresh water to create hydrogen to run the gas plant.

Hydrogen is discussed in a chapter of its own, but the arrangement has obvious, major inefficiencies. The renewable energy plant is the main energy source, so why not take power from that directly? Why bother with the hydrogen and gas plant? To have any chance of meeting the 30 per cent target consistently, the project will also need some so far undiscovered means of storing hydrogen safely in large enough quantities to tide the gas plant over long periods when the wind does not blow and the sun does not shine.

One of those who tried to convince the government, specifically the Federal Energy Minister Chris Bowen, that this eccentric approach just would not work was Snowy Hydro chief executive Paul Broad. As well as publically declaring that the commercial use of hydrogen as a fuel was years away, Broad also tried to tell the government that, with Liddell closing, the grid needed several peaking plants, not just one. Those plants could then be powered up very quickly when the wind dies over large areas of the eastern seaboard, which is expected to happen all too frequently, and turned off when it starts to blow again. This is not an efficient and certainly not a cheap way to run any grid, but activists will still have their wind farms and the lights will remain on, for now.

All that sensible advice seems to have fallen on deaf ears with Broad resigning in late August 2022, citing clashes with Energy Minister Bowen. However, the Liddell plant may be able to start operating using just natural gas, for now.

What all this comes down to is that Australian state and Federal governments are refusing to learn the lessons taught by now extensive operational experience with wind generation both in Australia and overseas. Wind and solar together simply cannot do the job of coal and gas plants, they have to be backed up by the despised fossil fuel plants.

Victorian government energy minister Lily D'Ambrosio toed a hard line on this matter by ruling out paying coal and gas companies to keep them operating as part of a proposed national capacity market. The minister claimed that the state's planned series of offshore wind projects, including ambitious targets of 2 GW by 2032, 4GW by 2035, and 9GW by 2040, will "blow any shortfall out of the water".

These offshore installations will, as far as anyone knows – none have actually been built yet – take advantage of the reliable trade winds which blow down Bass Strait. Well, are they reliable? The material on wind droughts cited earlier in the chapter should ring alarm bells but we can also look at what's happening on King Island, well out in Bass Strait. This has the King Island Renewable Energy Integration Project, part of which is a wind farm, plus solar power as a supplement to the island's long-standing diesel generators. Material produced by the owner Tasmanian Hydro estimates that renewable energy now accounts for 65 per cent of the island's power demand.

That's fine but what about the other 35 per cent supplied by diesel? Why couldn't the wind farm supply all the island's needs from the supposedly reliable Bass Straight winds, and was the outcome worth the $18 million spent on the project, all to service the island's 1,600 residents? The Victorian government could at least produce some material apart from activist assurances that its projected reliance on offshore wind farms will be anything but a disaster. The King Island project is discussed further in another chapter.

Offshore wind generators, incidentally, are a major engineering feat. They involve installing an immense concrete pylon on the sea floor and embedding a wind tower in it. The wind tower also has to be engineered to withstand major storms, rouge waves and the like. This is not cheap or easy. Floating wind turbines have been developed but these are even more expensive than fixed installations.

Activists, which includes some politicians, have proved blind and deaf to these and other problems with this push to build an all-renewables network simply from wind and solar assets. They deny the problems exist or repeat discredited assertions that building more wind generators will fix the problem or claim that the occasional blackout also happens with coal fired plants. A last ditch, dismissive response is to say that a few days without power is no big deal.

In the foreseeable future there is no way that wind power can replace coal or gas fired power. Australia's shift towards renewable energy promises disaster not salvation, with the loss of coal-fired capacity is likely to become critical when the Liddell power station closes in 2023.

Addenda – Western Australia

Grids in the Western half of the country, which are separate from the East Coast grid, have received little attention in this narrative so far, mostly because the State Labor government of Mark McGowan has not proved as fanatically determined to wreck the future of its residents as state governments in the East.

A report issued in mid-June by the AEMO which also has some management oversight of the Western grids, notes that there will be a minor shortfall in generating capacity after the Muja C coal plant closes in 2024, but the state should be fine until 2025. There are two other small coal plants and a surprising number of gas power plants of one sort of another in WA, which may help explain the ability of South West Interconnected System (to give the grid is formal name) to cope with an enormous number of roof top solar installations.

These photovoltaic panels have been known to generate 78 per cent of the grid's supply at midday on a sunny but otherwise mild day. Grid operators can turn off gas plants when the PVs are delivering power, to balance supply to the grid, and turn them back on again in the late afternoon, in time for the evening peak. Rooftop solar is of no use at night. Coal plants cannot be turned on and off in that fashion.

Because the state government and activists have not been messing with major sections of the generating capacity of the state, power price increases are not nearly as bad for WA residents as they have proved in the Eastern half of the country. In the third quarter of 2022 (quarter ending September) wholesale power prices were less than a third of those in the National Energy Market.

4

POOR AND POWERLESS

* When brown coal power stations ruled the state-owned grids, power was cheap.

* Even before the crisis of 2022 prices were trending up, with the main factors being the general aging of conventional power plants and increasing prices for fossil fuels.

* Renewable generators are, at best, some help in reducing prices in certain circumstances. However, so much additional storage, transmission lines and additional generators have to be added that if such generators are used in substantial numbers, they become an expensive drag on power grids.

* Indications from overseas grids which have considerably more experience in dealing with renewables is that they increase prices wherever they are used.

An illustration of the problem that consumers now face in power prices is the experience of Peter Feros, a member of the Energy Realists group mentioned in the previous chapter, when he asked for quotes for the supply of electricity to a small, historic hotel his family owns North of Port Macquarie in NSW. The best offer on renewal of the contract worked out to an annual increase of about $25,000 – a price hike of almost 60 per cent.

The cheapest bidder, energy supplier AGL, offered a range of explanations for the increase including power plant closures, the Russian-Ukrainian war affecting energy supply and wet weather also hampering coal production and delivery.

Similar explanations would be offered to businesses renewing power contracts all over Eastern Australia, but whatever the reasons given they would still have the higher bills and pass those additional costs onto their customers. Families slugged with those higher prices, not to mention their own higher power bills, will have to reduce spending in other areas.

All of this, in turn, is because a group of comparatively well off activists and members of the political elites decided that the reliable coal-fired power plants had to go.

For the current surge in prices is not so much due to renewables, although they haven't helped, but to the relentless demonisation of all things concerned with fossil fuels. Around the world new coal and gas projects and fossil fuel power stations have been delayed by endless legal actions and demonstrations. Major financiers have been lobbied, cajoled and bullied into giving up lending to such projects, and governments attacked until they take extreme action such as banning new gas projects, or even the search for new deposits of gas.

In Australia, this relentless campaigning backed up a sense of urgency unjustified by anything that is actually happening in the environment means that no new coal powered plants have been built for years. As discussed in the previous chapter, the one major gas power plant now being built is due to government action. At the same time activism is forcing reliable black and brown coal plants to close at an ever-increasing rate, usually before the end of their service life. The Australian Energy Market Operator which runs the grid for the Eastern half of the country estimates that 60 per cent of the grid's capacity will close by 2030, but that estimate was made before announcements in late 2022 concerning further plant closures.

As a result, and as discussed in earlier chapters, power consumers are being left at the mercy of an aging fleet of coal plants which are increasingly going off line for one reason or another, a network of erratic renewables that may or may not decide to deliver power when the coal plants are offline, and skyrocketing coal and gas prices.

Even before the 2022 crisis, electricity prices were generally trending up. A graph in the annual retail markets report for 2020-21 produced by the Australian Energy Regulator shows that from 2005 to 2018 retail electricity prices climbed to be about 80 per cent plus above 2005 prices in real terms (that is after adjusting for inflation). Prices declined from that peak but even after that decline in 2021 electricity still cost 60 per cent more than it did in 2005, while incomes have increased just 20 per cent.

Then the crisis of June 2022 hit, and it was an acute one. As noted, commodity prices were trending up before that, thanks to the tireless work of activists, but Russia's invasion of Ukraine early in 2022 pushed prices to a tipping point. By failing to develop its own gas reserves and rejecting other sources of energy, such as Germany shutting down its nuclear plants, Europe was too heavily dependent on gas supplies from Russia which were badly disrupted by the war.

In Australia, the result was a perfect storm of aging coal plants suddenly off line all at once, sky rocketing fossil fuel prices and renewables failing to deliver. The crisis was particularly acute in Queensland with advisory company Energy Edge noting that prices for wholesale power in the state more than doubled to an unheard-of average of $323 a megawatt-hour in the June quarter. Overall wholesale prices for the Eastern grid hit $216 a MWh and were still at about that level in September. When brown coal fired power stations ruled the old state grids 20 years ago, wholesale power might have cost about $40 a megawatt hour. Just a few years ago, the price was $80 a megawatt hour.

This will obviously have a major effect on prices paid by consumers, and that price pain is set to get worse. In a budget statement delivered in October 2022, Treasurer Jim Chalmers revealed households could expect a 56 per cent increase in power bills over the next two years. Power bills would rise 20 per cent in the second half of 2022 and a further 30 per cent in 2023-24 (these estimates have since been pegged back).

To return for a moment to the pre-crisis rise and slight decline, the reasons for this price behaviour are complex. Electricity bills paid by consumers are typically roughly evenly split between wholesale power prices and the costs of building and maintaining a power grid, with a slice of perhaps 10-20 per cent left for administration costs and profits. Part of the price increase up to 2018 was the vast investment required to remake the aging grids left over the days of state-run networks and meet new technical requirements. Part of the decline since then has been due to the remake being complete. On the generator side, reliable coal power stations were falling out of the market and not being replaced, although for some years that did not matter as there was considerable over-supply in the market. As noted, renewables play a comparatively minor part in this story, but they have helped push reliable plants out of the market by reducing their profit margins. Although they wholly fail to substitute for the reliable power, when the wind blows they can dominate the spot market, taking business from coal. Crucially, there is no capacity market in Australia, where generators are paid to be ready to supply power. More on this in a moment.

The Australian Energy Regulator's Wholesale Markets Quarterly report for the fourth quarter of 2021 points to other factors such as generally milder summers reducing demand (global warming theory says summers should be getting hotter not colder), plus the growth in rooftop solar.

Activists typically ignore all the complexities of the debate and seize on the decline since 2018 as "proof" that renewables reduce the cost of electricity. They have had to work harder to explain away the sharp reversal in that decline by the previously mentioned crisis, but there is at least some justification for their excuses. Power prices have increased due to a spike in energy prices, rather than anything to do with renewables as such, plus a continuing decline in conventional power capacity.

Here we need to clarify a few points. Renewables by themselves can offer power often at much cheaper prices then the derided conventional power plants, when the weather conditions are favourable. In fact, spot prices (that is, current market prices as opposed to prices set under fixed term contracts) have been known to fall to zero and even go negative (the customer is paid to take the power) during favourable conditions, as wind and solar installations deliver more power than the market can handle. There are also comparisons called levelised cost estimates produced by various bodies which supposedly shows that power generated by wind and the sun is cheaper than that of coal fired plants. These were never intended to compare the costs of using dispatchable power (coal, gas, nuclear) with those of intermittent sources (wind and solar) on a grid. The costs of using such power sources to produce electricity 24/7 on a major grid are an entirely different matter.

As is discussed in another chapter, the trouble starts when the wind dies down and the sun is not shining. Energy stored in batteries and in pumped hydro projects such as the much hyped Snowy 2.0 – when it is finished and assuming it can find enough fresh water to operate properly – will tide the network over for a time but at some point the despised fossil fuel generators have to be switched on, or a lot of angry power consumers will be left in the dark, and then asked to pay huge power bills.

As noted, the closure of the reliable plants did not matter for a few years as there was considerable over-supply in the market. Now the market is beginning to show signs of real wear around the edges and the effect of capacity closure on prices is marked.

When Victoria's Hazelwood power station representing 1.6 GW worth of capacity closed in 2017, the Australia Energy Regulator later noted that average electricity spot prices increased 85 per cent year on year in Victoria, 32 per cent in South Australia and 63 per cent in New South Wales and Queensland.

Not only are prices starting to increase the grid is becoming more erratic. As readers will recall, it is not enough to just cover output, the system must have a certain amount of capacity in reserve for emergencies and that's proving a problem.

This point was made by the Australian Energy Council in a submission on a focus paper on Wholesale Market Performance Monitoring issued by the Australian Energy Regulator late in 2021. The submission says that before 2017 (that is, before Hazelwood closed) the AEMO rarely issued what are known as intervention orders. Now they are commonplace.

These orders may involve directing a gas generator to remain in the market, or a diesel plant to continue operating because the AEMO has realised that there is not enough reserve capacity. The AEC, which represents the biggest operators in the energy market, says that these orders have become seemingly the default way of managing the market.

The last people to acknowledge these problems are the activists. But at least the crisis of June 2022 finally forced policy makers to confront the need for reliable power supply with governments groping their way towards the formation of a capacity market, where generators are paid to be on standby for when the wind fails and night falls. Such arrangements are so common overseas that commentators have noted an increasing disconnect between spot market prices and actual prices paid which includes both set contracts and payments for capacity. As this book has been written staunch resistance from the Victorian government among others, due to the demonisation of fossil fuels, has blocked the development of a capacity market.

Instead the Australia grid is being dragged towards a renewables future although, as noted they have yet to do much direct damage to prices. In certain circumstances it is even possible for the many staunch defenders of renewables to claim that they have reduced prices (sure, from a peak). They also claim that the equipment is getting cheaper which may also be true, but as we have seen there is a big difference between a standalone installation generating power, and one which has to plug into a grid designed to deliver power 24/7 adjusted to suit the needs of those using the grid.

Not only does the grid need enough of these renewable generators to supply power at any given moment, as discussed it needs:

* A massive amount of expensive storage to tide the grid over times when the wind does not blow and the sun does not shine.

* Major additional transmission lines and interconnectors to link the often remote areas where the renewable generators have been set up.

* A massive overbuilding of generators in all areas to compensate for the intermittent lack of wind and sun.

* A stream of replacement equipment. PV panels and wind generators have comparatively short working lives. Perhaps 25 years or so for wind turbines and 20 years for PV panels. A large number of these installations would then require a continuous stream of replacements.

This does not sound cheap at all, or very green (the problem of disposal of PV panels and wind generators is discussed later in the book). Nor is it possible to point to any grid where the increased use of renewables unequivocally resulted in cheaper prices for power. In fact, the opposite is the case.

Higher power prices in overseas markets can also be attributed to high commodity prices, but before the crisis of 2022 there was a distinct correlation between expensive electricity and increased use of renewables. Electricity prices for EU countries compiled by the union's statistical body Eurostat for the first half of 2021 found that green-mad Germany had the region's highest electricity prices, followed by Denmark which has been working hard to decarbonise its power supply, and Belgium where green regulations are forcing power companies to buy carbon credits. Next on the list are Ireland, where wind farms produce more than 35 per cent of the country's power, and Spain which is also renewables mad. In contrast, Poland (where coal is still king) and nuclear-loving Hungary and Bulgaria, rank well down the list.

Having left the union, the equally green mad UK is not listed but a survey conducted by a boiler and air conditioner installation group called Boxt concluded that the UK had the highest electricity prices in the world. Although Boxt obviously has an interest in selling energy efficient equipment, high power prices are now causing severe political headaches for the ruling Tories. Australia ranked sixth on the group's list, slightly ahead of Switzerland and above even Germany.

The US is no different. A 2020 report by the US Energy Information Administration lists looney-woke California as well as Massachusetts, where wind farms rule, as having notably higher power prices than the rest of the country. A major exception is that of staunchly Republican Texas, where farmers have developed a profitable sideline in leasing wind swept broad acreages to wind generators. These export power to other states looking to make up green energy quotas.

There are economists who happily reanalyse all of this to show that the increases are all the result of additional taxes on power prices or other local conditions. However, it is clear that renewables, at best, don't help much, and may be making things much worse by drastically reducing the reliability of the grid.

This may be the trade-off. Power prices may come down but only if consumers are prepared to accept periods of varying length when they don't get power or have to make their own. Sales of small generators suitable for homes have soared in Texas and California, both states where residents were left without power for some time – for days in the case of Texas (although that was not due to renewables as such). The Californian government responded to this by banning the sale of small generators in the state. Californians who still wanted generators then went across state lines to buy them.

Those who don't take such action may eventually pay less for power but only if they are happy to live in dark, cold homes for lengthy periods, listing to activists on battery powered radios telling them that renewables are the answer to everything.

5

NET ZERO NONSENSE

- There is no real possibility of the world achieving net zero emissions. Perhaps the rate of growth in emissions might be slowed but that seems hopeful.
- Renewable energy is a partial, expensive and unsatisfactory solution to the problem of emissions from the electricity sector, but technologies for doing the same in steel and cement making are still in development and are likely to be costly.
- There is no technological way of cutting emissions completely in freight transport, sea transport and airline travel.
- Despite all the urging and declarations of imminent climate doom, the world's overall consumption of fossil fuels has barely been affected.
- In any case few countries are showing any real interest in meeting even their self-imposed targets under the Paris agreement, which are often trivial or token targets, let alone the immense effort that would be required for net zero.
- Declarations by this and that country leader that they will strive for net zero by some date far in the future amount to so much hot air.

The Paris treaty hammered out in 2015 and which came into effect in 2020 permits each country to nominate its own emission goals, called Nationally Determined Contributions

(NDCs). These can vary widely and for many countries, notably top emitters India, China and Russia, the NDCs were set so that those countries did not have to do anything at all. There is a country by country listing of top emitters at the end of this chapter.

In any case, the goals themselves are not legally binding. The only legal obligation is to nominate goals, and to lodge updated, improved goals every five years or so. The improved or tightened goals lodged to date have proved of little more use in limiting emissions than the promises made in Paris.

In 2019, Climate Action Tracker, a group run by three climate organisations, looked at 32 countries which collectively account for 80 per cent of greenhouse gas emissions, to find that by the group's criteria just seven were doing enough to meet the Paris goals. Only one of those countries, India, was in the top ten of CO2 producing countries. However, CAT's endorsement of India was basically wishful thinking. The group was impressed by the fact that the country had already met a major Paris goal of 40 per cent of its installed electricity base being non-CO2 generating plants, including hydro, biomass and nuclear as well as wind and solar. But the country was already almost at that figure before Paris, and the figure is for "installed" base. For solar and wind that can be very different from actual power generated.

Despite this evident lack of action at some point activists lifted expectations by talking about net zero emissions, which is a whole world of pain away from the Paris agreement.

All the talk about net zero prompted the International Energy Agency to release a report *Net Zero by 2050 – A Roadmap for the Global Energy Sector* which found that the goal could be achieved, albeit only if all economies followed a "narrow but still achievable" path. But that narrow path, which included an immediate doubling in annual investment in the energy sector with no money at all going into the fossil fuel projects, requires commitment and dedication to the net zero goal which very few countries have shown to date. (The report effectively meant that net zero was never going to be achieved, but the agency was understandably reluctant to make that stark assessment.)

Two sticking points, among many others noted in the report, are the need to reduce emissions from both the steel and cement industries. The IEA report admits that technologies that might reduce emissions in those industries are still in demonstration phase, at best. Uncritical reporting on various supposed green hydrogen breakthroughs never says anything about the likely costs of these processes, versus the tried and tested ways of producing energy, steel and cement, but it is likely to be high. One estimate put the additional cost of producing green steel at about $US70 a tonne, which could cause a real problem. If Western producers adopt this technology and the Chinese don't bother (a very likely scenario), then all that happens is that the Chinese produce more steel at prices that undercut those of their western competitors, and emissions remain the same.

Freight transport, by truck or ship, and airline travel is rarely mentioned in the net zero debate because there are no effective ways to reduce emissions in those sectors. Better to buy expensive emissions offsets and pass the costs onto the consumers.

Another report produced by the IEA in May 2021, *The Role of Critical Minerals in Clean Energy Transitions* is even less reassuring as it points out that wind farms, photovoltaic panels and batteries require a great deal more of materials such as nickel, rare earths, manganese and chromium that were previously only in niche products. Depending on the material, the report notes, the required mines may take up to two decades to develop. Again, the basic message is that net zero is not going to happen, the IEA just does not want to say that. The problem of sourcing these materials is discussed further in the chapter on electric cars.

Net zero enthusiasts who want further discouragement could

always look at a control panel for the National Energy Market (the Eastern Australian grid) compiled by the Australian Energy Market Operator. This shows that in the 12 months up to around January of 2023 just short of 70 per cent of electricity came from black and brown coal plants, and just short of 20 per cent from solar and wind. Another 7 per cent came from hydro (which counts as a renewable) and a few per cent from gas.

This does not seem very different from the energy mix of preceding years, but the really bad news for activists is the total energy mix figures for Australia compiled by the International Energy Agency. This analysis adds in the use of fuel in domestic and freight transport, gas for cooking and industrial use plus the power required for electrical generation. In 2021, despite all the talk about net zero emissions, wind, solar and biomass collectively amounted to just a few per cent of the total energy task.

The story is the same for the world statistics also compiled by the International Energy Agency. These generally attribute about one third of energy consumption to oil, to run cars and trucks, with about half that going to freight transport. Gas and coal together add up to about half. That leaves a little more than 15 per cent coming from low-carbon sources, but a bit more than 10 percentage points of that is from nuclear and hydro power. Wind, solar and biomass (burning wood and sugar cane waste) then account for just 5 per cent or so of the world's total energy consumption. This is increasing but off a very low base and a glance at the figures for the past decade or so show that those energy sources are not increasing at anything like the rate to achieve net zero in the foreseeable future, if at all.

A conclusion that can be drawn from all of this is that the increasing talk about net zero comes from those who simply have no idea what it means or what it entails and ignores a history of many countries making token efforts only to reduce emissions. Declaring support for the concept of net zero, however, has certain conveniences for those who make such declarations. They can bask in the praise of an uncritical media knowing that the usual deadline of 2050, or even 2030, is so far beyond the usual media cycle that they are never going to be held to account for the promise.

List of top emitters (share of CO2 produced annually)

China (28%) – President Xi Jinping's announcement in September of 2020 that China would aim for peak emissions by 2030 and carbon neutrality by 2060 would have been more believable if country had not been, at the same time, binge-building coal fired power plants. This binge building is still going on at the time of writing, but capacity under construction is far more than the Chinese economy will need in the short term. Instead, they are being built with loans from the central government to stimulate the economy in certain provinces. Climate just isn't relevant.

Thanks to the energy crisis there are also signs that President Xi is now unwilling to play the climate game. In his speech to the 2022 Party Congress, President Xi, in effect, declared that the country would not stop burning fossil fuels until it was confident that clean energy could reliably replace them.

"Based on China's energy and resource endowments, we will advance initiatives to reach peak carbon emissions in a well-planned and phased way, in line with the principle of getting the new before discarding the old," President Xi declared. A work report released after the speech states that China will expand exploration and development of oil and gas resources and increase reserves and production as part of the measures to ensure energy security.

The country is also spending a lot of money on wind generators and solar farms, probably for much the same reason as it has been building more coal power plants. A few years ago the Chinese were building dams everywhere. Before that it was smelters, conference centres, and, oh yes, whole cities in which no one lives. Net zero emissions hardly seems relevant.

China's attitude to climate alone completely destroys any prospect of the world reaching net zero, or even of reducing the growth in emissions. One glimmer of hope, however, for those who are determined to destroy the world economy in order to save it, are the endless warnings about the health of the Chinese economy. If these prove correct and the country's economy collapses there will be fewer emissions. As commentators have been warning that China is on the brink of collapse for well over a decade now, activists should not hold their breath waiting.

USA (15%) – American President Barack Obama signed the 2015 Paris treaty as a presidential agreement, not as a treaty binding the country. The wording on the document was even changed so that it could not be considered a treaty under American law. Otherwise, it would require a two thirds majority in the US Senate to be ratified (that is, accepted into law), and that's never going to happen. Some years down the track President Biden has a climate agenda to match his rhetoric and has managed to get a massive package called the Inflation Reduction Act through both legislative houses. This pledges US$369 billion in climate investments over the next decade which, scientists hope, will cut US greenhouse-gas emissions by about 30–40 per cent below 2005 levels by 2030. There is a strong element of wishful thinking in these estimates. The biggest chunk of money from the legislation is $US128 billion in tax credits over the next decade for businesses shifting to greener power sources, such as solar, and individuals installing heat pumps and the like. The UK's experience with attempts to get consumers to install heat pumps as a means of reducing emissions has been far from a success, and businesses will want to do the least possible work to claim the credit. In the meantime, the states, including the likes of loony-woke California and staunch Republican Texas, are all pursuing their own wildly different agendas. California intends to ban the sale of petrol driven cars by 2035, and the state of Wyoming has declared (perhaps not seriously) that it will ban sales of electric cars by the same date.

India (7%) – as noted India set up its Paris pledge so that the country can meet it easily. Another part of its pledge was to "cut greenhouse gas emissions intensity of its gross domestic product by 33% to 35% by 2030", but this was meaningless as the intensity was declining anyway. As India remains a developing country with a power grid so rickety and blackout-prone that most major users have their own diesel generators, a declaration about net zero emissions would not mean much in any case. For a time after Paris India had its hands full with the Covid pandemic, with no time to spare for emissions reduction. Covid may have passed but, like all developing countries, India is more interested in a $US100 billion a year fund which the rich nations are supposed to create in order to help the poorer nations reduce emissions, than in actually reducing emissions. Give us some money, the country is saying, and we will think about reducing emissions.

The latest climate meeting in Egypt resulted in a proposal for a one trillion dollar fund by which rich nations pay compensation to poor nations for damage due to climate, details to be worked out later. This was hailed as a major breakthrough but probably means about as much as the earlier proposal for the $US100 billion a year fund, which to date has attracted only a tiny fraction of the proposed amount. These proposals are more about making those who come up with such schemes look good, than about solving anything.

Russia (5%) – Now an international pariah thanks to its invasion of the Ukraine, Russia is not going to be interested in aiding the West in its obsessions over emissions. In any case, the country set up its pledge to the Paris treaty so that it did not have to do anything to meet its goal. With its economy wrecked thanks to the war it still won't have to do anything to meet climate targets, assuming it can be bothered setting them, but the fighting is generating plenty of emissions.

Japan (3%) – In October 2020 Prime Minister Yoshihide Suga declared that Japan would achieve carbon neutrality by 2050. Fine words but the current Paris target is for a 26 per cent reduction in emissions from 2013 levels by 2030, and news outlet Nikkei Asia notes that the country still has 150 coal fired power plants. The government is pushing for the efficiency of the plants to be increased which may be useful but hardly sounds like the drastic action required to achieve net zero. Japan has previously shown little interest in renewables but occasionally announces stuff like new offshore wind farms and has dabbled in Hydrogen.

Germany (2%) – Like the UK, Germany is one of the few countries that has shown any political will in cutting emissions to the point of spending billions of Euros on solar panels, although it is not a naturally sunny place. After previously declaring that it will phase out coal fired plants by 2038, the country brought that date forward to 2030. Now thanks to its supply of Russian gas being cut off due to the war in the Ukraine, the country is turning its coal plants back on and scrabbling for any new source of energy it can find to the point of bulldozing a village in Western Germany to make way for a new mine. In any case, even if wind turbines and solar panels can fully substitute for coal plants, which they can't, it's not clear how anything like enough turbines could be built in time to meet the previous, ambitious targets for phasing out the country's still active brown coal plants. All the best spots for wind turbines, offshore and onshore, have been taken and new projects are encountering increasingly tough opposition from nearby residents. (The same activists who want zero emissions often campaign against transmission lines and wind farms in their areas.) All this effort and spending has not had much affect. There have been reductions in emissions, but this may be more due to industry relocating to China.

Iran (2%) – One of the few countries not to have ratified the

Paris treaty (officially adopted the treaty, as opposed to signed), the country has many other matters to contend with besides emission reductions.

South Korea (2%) – This country has declared that it will cut emissions by 24 per cent from a 2017 baseline by 2030. However, it plans to do so by buying green credits from overseas and, according to Climate Action Tracker, by increasing the carbon absorbed by forestry and land use. Experienced observers of the climate action scene suspect that the country may simply fiddle with assumptions for carbon absorbed by forests and land use to find extra credits, as other countries did to meet obligations under the earlier Kyoto protocol. No need to bother industry or voters.

Saudi Arabia (2%) – The goal declared under the Paris treaty involved cutting emissions and increasing the use of nuclear energy and renewables but this was made dependent on "a robust contribution from oil export revenues to the national economy". Oil prices have varied quite a bit since then and the kingdom has declared ambitious goals for renewable energy. However, as less than 1 per cent of the country's power now comes from such sources there is a long way to go.

Indonesia (2%) – Listed as a major emitter because of its treatment of forests and carbon-rich peatlands the country has promised to clean up its practices in this area, which includes doing something about slash and burn farming. Farmers will clear an area for food production, by cutting the trees on it and then burning the area. The resulting ash, rich in nutrients, fertilises the land nicely and the fire gets rid of many of the pests. The land is then good for three to five years and, when it stops being fertile, the farmers move on to another area, leaving the jungle to regrow on the land. The cycle is later repeated. This form of primitive, subsistence farming can cause major fires, as occurred in 2019 – fires which produce more carbon dioxide per day than all of American industry. Ending the practice is easier said than done as it occurs in areas where the government often has little control. These farmers don't usually start the day by consulting government websites to find out what they should be doing.

Other notes

The UK – not one of the top ten emitters but it is a developed nation that, before the energy crisis, was trying hard to reduce emissions. When Boris Johnston was still in power the UK has announced all sort of initiatives for reducing emissions including building a host of offshore wind farms (finding suitable sites onshore is now too hard) as well as proposing that anyone selling their house has to replace the standard boiler with a less efficient heat pump.

The heat pump proposal is worth noting as such pumps were expected to cost more than £10,000 ($A18,400) per house, with the UK media reporting that installation may cost more than three times that, depending on what has to be done to the wiring and pipes of old homes. The proposal seems to have fallen by the wayside, but it is the sort of drastic, electorally unpopular action that must be taken to have any hope at all of getting to net zero in the foreseeable future. Few countries are going to take that road.

European Union – the EU is often considered one entity for any listing of emissions, ranking about fourth in the world. Collectively the union also has one of the strongest climate policies. In December 2020, EU leaders endorsed a binding EU target for a net domestic reduction of at least 55 per cent in greenhouse gas emissions by 2030 compared to 1990. This sounds good but a look at the attachments to the original announcement reveals gems like this, "some hard to abate sub-sectors, notably aviation, will also require the development of advanced biofuels and sustainable alternative low or zero carbon fuels and gases". There is also a discussion on re-evaluating the EUs "land use sink", in other words "discovering" that more carbon has been absorbed by Europe's forests than previously calculated.

All that said the EU finally has a carbon price through its Emissions Trading Scheme that is high enough to affect change. After years of struggling to get above €20 ($A30.88) per tonne and plagued by instability, the price of abating a tonne of CO_2 exceeded €90 just before Russia's invasion of Ukraine. The invasion caused the price to fall to the high €60s, which is still pretty good. At the end of 2022 the price was €75-80 However, a high carbon price means that voters will be digging deeper into their own pockets to pay for climate change, at a time when energy prices are already very high. They will not be happy.

The energy crisis and sky high energy prices brought on in part by the war in the Ukraine has since prompted a major reconsideration of climate and energy goals throughout Europe. It seems that voters are more interested in not freezing to death because they can't afford to turn on their heaters, than in climate.

6

HYDROGEN IS NOT THE NEW LNG

* The use of hydrogen as the medium of a power export market has an obvious, major flaw. Unlike coal or gas, hydrogen can be created anywhere where there is water, wind and sun. Why will any country import the gas when they can make it on their own territory?

* Hydrogen is not like LNG. It is much harder to put into liquid form, is much more likely to leak and has different properties which make it a far more dangerous gas.

* Hydrogen has been used as a feedstock in many industrial processes for decades, but the vast bulk of the gas is consumed in the same place it is made, from methane and steam – this is cheaper than manufacturing the gas by using electricity.

* Energy losses from converting electricity generated by renewables into hydrogen and then back again at the other end means it is far more wasteful than a transmission line, so why not use a transmission line instead? A battery is also a more efficient and safer means of storing power, at least compared to hydrogen.

* Hydrogen's main use would seem to be is as a comforting

fantasy for activists to tell one another.

If the author of this book had to hand out awards for the worst idea among all the proposals for generating and storing green energy, then the mass use of hydrogen as a sort of alternative to Liquid Natural Gas would be a major contender for the top prize.

Unlike power from coal and gas renewable energy can be generated anywhere, and almost any country that can be named has at some point talked about becoming the "Saudi Arabia of wind" as UK Prime Minister Boris Johnson put it. In other words, why would say, Japan, import horrifically expensive power from elsewhere when they can make horrifically expensive power in their own territory, including coastal waters?

This point was forcefully made by a Professor of Engineering at the Australian National University, Andrew Blakers, in the Australian edition of The Conversation, an online site for academic articles, in April 2022. An enthusiastic and tireless advocate for renewable energy Professor Blakers says that the federal government has tried to encourage a hydrogen industry by setting aside hundreds of millions of dollars to expand Australia's green hydrogen capabilities. Those efforts will include creating a major green hydrogen export industry, particularly to Japan, for which Australia signed an export deal in January.

However, he also points out that Japan has more than enough solar and wind energy to be self-sufficient in energy – assuming all that energy can be harnessed. Whether or not you agree with Professor Blakers that Japan can realistically meet all of its energy needs from local renewable energy the country can certainly generate hydrogen locally.

Some background: hydrogen is currently used as a feedstock for many industrial processes such as treating metals, producing fertilizer, and processing foods. Petroleum refineries use hydrogen to lower the sulphur content of fuels. Almost all of that commercial hydrogen comes from the traditional method which uses steam and natural gas, as that is by far the cheapest way of extracting hydrogen.

Proponents of renewable energy, however, now want to build hectares upon hectares of wind farms and solar energy generators to make hydrogen by electrolysis – that is, by passing an electric current through water. The idea is to store this hydrogen in some way, preferably in liquid form like LNG, then ship it off to where it is needed as a replacement for fossil fuels in generating electricity, to power cars, and perhaps even ships and planes. This is basically the vision set out in a 2019 report produced by the impressively named Council of Australian Governments Energy Council Hydrogen Working Group, chaired by Australia's chief scientist of the time, Professor Alan Finkel.

However, as was also mentioned in the Finkel report, although perhaps not prominently enough, the process of making, condensing and shipping hydrogen is known to be technically challenging and wasteful. Hydrogen powered devices are also less efficient than the fossil fuel versions.

Professor Blakers cites an estimate that converting energy to hydrogen, shipping it to where it is needed and then converting back into energy could consume 70 per cent of the energy generated. Michael Liebreich, a senior contributor to BloombergNEF (New Energy Finance) wrote in 2020 that as an energy storage medium, hydrogen has only a 50 per cent round-trip efficiency – far worse than batteries. He estimated that hydrogen powered fuel cells, turbines and engines are only 60 per cent efficient – far worse than electric motors – and far more complex. As a source of heat hydrogen costs four times as much as natural gas. As a way of transporting energy, hydrogen pipelines cost three times as much as power lines, and the cost of sending the gas by ship of trucks is even worse, he says.

Activists who talk so glibly about using hydrogen to store energy are no doubt thinking of Liquid Natural Gas, which is the subject of a thriving international trade using purpose-built container vessels. However, the technical problems of shipping LNG were worked out, the facilities were built, and customers were found to buy the output before the general public was fully aware of the general usefulness of being able to trade gas across oceans.

Unlike LNG, hydrogen presents considerable difficulties in its storage and use. It is a much smaller molecule than methane, so seals and pipes that would comfortably prevent methane leakage do not keep hydrogen in. The liquification temperature for hydrogen is much lower than that of methane, specifically minus 253 degrees centigrade or just 14 degrees above what physicists call absolute zero – you can't get any colder – and so requires considerably more energy to achieve and maintain. The alternative is to store the gas under very high pressure.

This leads to the problem of safety. Without getting into technical details, hydrogen has different burning and explosive properties to that of LNG and, as noted, a greater tendency to leak. It is a far more dangerous substance. History buffs will recall the 1937 explosion and fire that destroyed the German airship the Hindenburg, which used hydrogen to stay afloat. The technology of airships for passenger transport was abandoned after that, but a few such aircraft remain in service by using the much less dangerous helium rather than hydrogen to stay aloft. At the very least, major hydrogen systems will require a stringent set of safety rules and procedures which may have to be learned the hard way.

Then there is the problem that switching to hydrogen is not just about slapping a hydrogen tank on an existing engine or using existing pipelines. Everything will have to be redesigned and rebuilt, all at eye-watering cost.

Faced with these inconvenient facts, activists offer counter arguments that range from the feeble to the ridiculous. They claim that green power will be so cheap the wastage from using hydrogen to store the power will not matter. Really? Refer to the chapters in this book on renewable energy. In any case if the power is so cheap why wouldn't each country create its own power and never mind any export market? If energy has to be shifted around internally, why not reduce the losses by using a transmission line? If power has to be stored then massed batteries may be almost as ridiculous a solution, but at least it would be cheaper, more efficient and (probably) safer than a hydrogen storage unit.

Another argument is that hydrogen can be stored cheaply in salt domes. These geological features are a key part of the formation oil deposits. The salt can be extracted comparatively easily to form large, underground pockets for gas storage, or so it is hoped. There are development projects in the Europe and in the US looking at salt domes but the last word in this area such be left to another BloombergNEF report.

"Storing hydrogen in large quantities will be one of the most significant challenges for a future hydrogen economy. Low cost, large-scale options like salt caverns are geographically limited, and the cost of using alternative liquid storage technologies is often greater than the cost of producing hydrogen in the first place."

Activists also point to hydrogen's possible use in town gas supplies, which occurred in the 1960s before the advent of bulk trade in LNG. Gas appliances can be used with a certain portion of hydrogen in their fuel, perhaps up to 30 per cent. Anything more than that and both the appliances and gas mains will have to be rebuilt.

There are already niche uses where the advantages of hydrogen outweigh the disadvantages such as in rocket fuel and fuel cells for submarines and so on. Otherwise, it is simply too difficult and costly to use without further technological breakthroughs.

Although there have been reports of major deals involving hydrogen, it is clear that hype has outrun reality. In fact, to judge by the large amount of nonsense spoken and written about its use, the main value of hydrogen is not commercial at all. The gas's main value is to provide comfort to activists. It is one of the many fantasy stories they tell themselves in the expectation of some day reaching green nirvana, somewhere over the rainbow. It is about as much use as any other fantasy story.

ADDENDA

People who think they know stuff will lecture others at parties about the exciting possibilities being opened up by a proposal to build an undersea power cable between an array of renewable energy projects in Australia's top end and Singapore. But the proposal runs into the same objections as trading in Hydrogen.

Why would the grid authorities in Singapore and Malaysia buy power from this cable, which is estimated to cost $16 billion to build, as opposed to power from renewable projects close to hand? Sure, the area close to Darwin may have advantages in the production of renewable energy but renewable energy enthusiasts will also point to areas around Singapore, Malaysia and Indonesia (parts of which will be close to the planned route of the cable) where renewable energy can be generated, and there is better access to fresh water. Renewable energy enthusiasts are never short of ideas.

The question then becomes is there enough of an advantage in generating renewable power in Australia to justify spending $16 billion on a cable to ship it offshore, plus a lot more to build renewable energy projects and batteries? Whatever the answer to that question might be, a basic requirement for the project should be guarantees from the Indonesian, Singaporean and

Malaysian grid authorities that they would buy enough power from the cable to justify building it.

The project, which has been in the public arena for some years, never lacks for positive publicity but all stories fail to mention any purchasing agreements. The Singaporean government has declared that it would prefer shipments of hydrogen, although this preference seems to be a part of a long term plan to research the technology rather than actually do anything about it. At the time of writing the founders of this company have fallen out over the company's need to raise another $60 million, and it has been placed in voluntary administration.

7

STORING UP TROUBLE

* Scientists have identified/discovered/developed plenty of ways of storing energy. There include some interesting ideas, but none seem viable for grid level storage for the foreseeable future.

* For now, we are stuck with batteries and pumped hydro projects as the only way to store appreciable amounts of power, and batteries are simply too expensive for grid level storage.

Scientists and activists have suggested solutions to the problem of storing energy ranging from the ridiculous to the practical but limited. These include hauling loads up hills. The energy is stored by hauling a load (anything heavy) up, say, a railway track on the side of a hill. The energy is released by letting the load slide back down. A version of this has been proposed for an old mine in NSW, using the mining shaft.

Other suggestions are for hydrogen or compressed air stored in underground caverns, liquid air stored in tanks, sand (a heap of it can be heated up and that heat is released over time), molten salt as a store of heat and different types of batteries, to name a few. Some of these proposals are technologically interesting, and all would work to some extent. However, they all run into the problems of cost and efficiency not to mention the sheer size of the task.

As noted in another chapter, if the wind stops blowing for, say 36 hours – this may happen at least once a year in the area covered by the National Energy Market (Eastern Australia) – the grid may need 900 GWh (gigawatt-hours) worth of storage to tide it over, and that would be a minimum figure.

The Hornsdale Power Reserve, the official name of the Elon Musk Big Battery in South Australia, now holds a little less than 200 MWh at a cost of $126 million. At that price just one GWh would cost more than $600 million. As powering the entire state for just one day may require perhaps 185 Hornsdale batteries, this form of storage would cost many billions of dollars to supply the grid for any length of time plus many millions of dollars more each year to continually replace batteries that wear out. To make matters worse grids are usually designed for worst case scenarios, of long periods of little wind and not much sun. Even more storage would be required.

Journalists occasionally claim that that battery projects make money, glossing over the fact that government support of one form or another is still required to build them. A survey by consultancy firm Wood Mackenzie found that, at the time of writing, Australian companies have plans to build 9.2 GWh of battery storage but only 4 per cent of these projects have started construction. In other words, although batteries are being built, total storage is far short of anything that would be regarded as significant at the grid level.

Batteries have at least one advantage is that they can be turned on in an instant when the wind fails or demand on the grid changes. The same can be said for about the only viable option for grid level storage, that of pumped hydro. This has been discussed in another chapter, but it should be noted here that unlike other hydro projects in Australia, including the original Snowy Mountains scheme, pumped hydro projects do not generate energy from a fast flowing river. Instead, they store water in upper reservoirs and, when energy is needed, let the water flow through turbines to a lower reservoir. The project then has to be recharged by the water being pumped back into the upper reservoirs.

The problem with this dam-as-a-battery approach is that once the water has run out of the upper reservoirs completely, it may take days to pump it all back, assuming there is excess energy on the grid to do this. The same objection applies to all the other dam projects suggested including dams on old mine sites, and one wild idea to build a salt water dam on the edge of the Gulf of St Vincent, using the waters of the gulf as the base reservoir.

A salt water dam should work as well as a fresh water one, but there are practical problems. The upper reservoir would be a massive pool of salt water drowning bush land and causing havoc with the water table and salinity levels in the surrounding areas. Nearby farmers would not be happy at all. A secondary problem would be building turbines that would work with corrosive sea water for any length of time.

Whatever their design dams cannot be built over night – the environmental approvals alone may take years to conclude – and Australia remains a comparatively flat, dry place with few areas suitable for hydropower. Academics have produced lists of sites where pumped hydro might be installed, without going into details about where the fresh water needed would come from or making any realistic assessment of likely difficulties in getting approvals.

Activists are getting rid of the fossil fuel power plants in favour of renewable generators as fast as they can, but the facilities to store power on the scale required to make up for the intermittent nature of green power are difficult to build, and that shortfall is unlikely to be resolved before any of the self-imposed deadlines for eliminated coal-powered plants. Until that problem is solved, however, Australian power grids have a choice between retaining substantial conventional power capacity or an unstable grid subjected to repeated brownouts and blackouts.

ADDENDA

For a time, the RE industry was enthusiastically spruiking various forms of renewable generation which would do the job of coal and gas fired plants by running through the night and producing power on demand.

One proposal in 2017 was to build a 135 MW solar tower in a field of mirrors in South Australia for $650 million. Known as Aurora, the project was based on a facility operating in the US which used molten salt (a combination of two chemicals, not table salt) to store enough energy during the day to hopefully keep running through the night. The proposal attracted the backing of the SA government and a lot of favourable publicity, all of which neglected to mention that the project required private financing to go ahead.

Investors declined to back the Australian project and the US company ceased operation in early 2020, having been unable to fix the problem of the salt leaking from its one operating facility.

There are other renewable generators which manage to operate around the clock, such as the 19.9 MW Gemasolar plant in Spain, which also uses molten salt and has been touted as the new wave of clean reliable energy. However, the plant, which started operating in 2011, has had few imitators. One possible reason for the slow adoption of this technology is that molten salt mixture is corrosive and so difficult to use, as the Aurora example seems to indicate. Special pipes are required. The material also has an alarming tendency to return to its natural, solid form at room temperature, meaning that it always has to be kept hot. After years of talk and development very few solar installations store power for longer than eight hours.

In late 2021 Dutch group Photon Energy announced that it would be developing a 300 MW solar plant in South Australia which will use water to store an impressive 3.6 GWh.

8

KILLING THE ENVIRONMENT TO CURE IT

* Batteries and wind turbines require materials to make, and those materials have to be mined.

* That means a vast increase in mining of material previously considered niche, in areas where unethical practices, such as the forced labor of children, is rampant.

* Boosting the supply of the materials to the levels required will take decades.

* Then there is the problem of disposing of the many thousands if not millions of devices such as wind generator blades and PV panels required to meet green energy goals, once they have reached the end of their useful life. This will present a major environmental burden for future generations.

The on-going so called transition to renewable energy and electric cars will have many strange results. For the devices activists spend so much time spruiking – wind turbines, batteries and solar panels – have to come from somewhere, and they require vast, additional supplies of minerals that were previously considered niche such as cobalt, lithium, graphite and rare earth minerals.

To take one example of the problems that can arise, about 70 per cent of the supply of Cobalt comes from the Democratic Republic of the Congo. An analysis produced by the non-partisan research group the Wilson Centre in September 2021 states that about one fifth of the region's production is small scale mining, involving people of all ages, including an estimated 40,000 children, some as young as six years. Much of the work is informal small-scale mining in which laborers earn less than $2 per day while using their own tools, primarily their hands.

Those driving electric cars may pause for long enough between green party meetings and wine tastings to respond that companies like Tesla have policies about responsible sourcing of materials for their components. A Fair Cobalt Alliance, formed in 2020, tries to ensure that the Cobalt used is mined ethically. The trouble is that about 60 per cent of the demand for the metal comes from the rechargeable battery industry in China which does not trouble itself about where the Cobalt has come from or under what conditions it is mined. No one ever accused the Chinese of being attentive to human rights. (Another point activists could raise in the defense of EVs is that manufacturers are switching to a new form of battery, a Lithium Iron Phosphate Battery, which does not use cobalt. However, that type of battery is not expected to overtake the use of batteries requiring Cobalt until 2028, when we will all be driving EVs, or so they say.)

China also has a chokehold on the production of rare earths, a group of 15 elements in the periodic table known as the Lanthanide series. Around 70 to 80 per cent of the world's supply of rare earths come from China. The rest comes from Australia.

Apart from the ethical issues, the problems of ramping production up to anything like the scale required to supply the transition are enormous. An International Energy Agency report released in 2021 *The Role of Critical Minerals in Clean Energy Transitions* notes that the production of Lithium graphite, cobalt, nickel and copper will have to be expanded enormously – assuming, of course, that countries take the Paris agreement seriously.

But even under scenarios assuming a much reduced rate of adoption of green technologies, boosting production to the levels required will take time. The IEA report estimates that it takes about 16 years on average for a newly discovered resource to reach production. Also, most of this development will occur in areas where there are few safeguards against environmental damage from poorly planned mining projects. Those buying materials for the green revolution may try to impose ethical standards on suppliers but, as the report notes, "it may be challenging for consumers to exclude" materials that do not meet environmental and social performance standards.

In mid-May 2022, The Australian Financial Review reported that the price of some battery materials has more than doubled in the previous 12 months. As a result, after declining since 2013, battery pack rises are set to increase in 2022 and remain high in 2023. The same article notes that Lithium prices have increased five-fold in just over a year.

Once the various items required for the green revolution have entered service they cause various environmental problems – wind generators are known for their ability to kill birds – but the real problems arise when they are discarded. Photovoltaic panels, for example, last perhaps 20 years out in the sun, wind and rain, before their performance degrades to the point where they are thrown away.

An article in the MT Technology Review in August 2021 (*Solar panels are a pain to recycle. These companies are trying to fix that*) says that about eight million metric tons of decommissioned solar panels could accumulate globally by 2030. By 2050, that number could reach 80 million. At the time of writing the US does not mandate recycling (although the EU does) and in the absence of any mandate most discarded panels go straight to land fill. The article points optimistically to companies working on ways to recycle these panels, focusing on the silver used to make them, but recycling panels remains expensive and difficult.

The problem is considerably worse for one of the main components of the wind farms now springing up everywhere – the gigantic blades. These blades are engineered to withstand hurricane force winds and, at the end of their life – maybe 15-20 years – they cannot easily be crushed, recycled or repurposed. All that can be done is to throw them into a landfill where they will remain forever. The statistics on the number of blades that have to be disposed of every year are mind boggling.

According to Bloomberg in the U.S. alone, about 8,000 will be removed in each of the next four years. Europe, which has been dealing with the problem longer, has about 3,800 coming down annually through at least 2022. It's going to get worse: most of the blades now going to landfill predate the major boom in wind farm construction.

All this adds up to a case of killing the environment in order to cure it. The quest for net zero, in itself an impossible goal, is set to have enormous, unintended, destructive consequences.

9

MESSING UP ON A SMALL SCALE

* There are plenty of micro-grids supplying small, isolated communities with power around the world that are partly renewable.

* Despite considerable effort and investment, none of these have achieved 100 per cent renewable production from wind and solar alone. The best achieved is 70-80 average renewable use.

* All these micro grids require some form of conventional back-up. For small grids this usually means diesel generators which can be powered up and down easily to meet major changes in renewable energy production.

* Power produced in this way is certainly not cheap, but at least it may be cheaper than solely diesel power, where the fuel has to be bought and shipped to a remote location.

In an ideal world governments and activists pushing for net zero would first construct scale model grids in which renewables take over electricity generation, hopefully at a reasonable cost. In fact, there are plenty of isolated, small grids called micro-grids such as El Hierro in the Spanish Canary Islands, Kodiak Island in Alaska and King Island in Australia's Bass Strait which make excellent test beds for activist ideas.

Unfortunately for activists, experience with wind and solar powered micro-grids simply underlines the fact that a fully renewables network using just wind and solar power is impossible. Some form of fossil fuel back up is essential. They also show that renewables can be ruinously expensive.

That said, the cost-benefit calculations occasionally favour the use of renewables in remote locations, as a means of cutting back on the consumption of diesel which is both expensive to buy and has to somehow be shipped to these locations. The real problems start when enthusiasts try to make a point by making the grids fully renewable.

Take El Hierro, the Western-most island in the Canary chain of islands off the coast of Africa and so well out into the Atlantic trade winds. Rather than import 40,000 barrels of diesel a year to supply the island's 11,000 inhabitants plus desalination plants with power, the island uses wind turbines backed by a pumped hydro facility, which cost €80 million ($A124 million) to build. This involves two fresh water reservoirs, one more than 700 metres up the side of the volcanic cone which makes up much of scenic El Hierro, and another near sea level.

Excess electricity from the wind generators is used to pump water up into the higher dam. When there is no wind or too much of it (wind generators usually do not operate during storms), the water is released from the upper reservoir through hydro turbines. This system can reach 100 per cent renewable generation on some days and has been known to operate for 25 days continuously solely on wind and solar power. As a result, the island has been hailed as a sort of poster child for our renewables future with various articles claiming, incorrectly, that the island runs solely on renewables.

A closer look at a page of production statistics compiled by the company which runs the installation, Gorona del Viento, kept firmly separate from the happy-PR talk page, shows that in 2020 about 42 per cent of the island's electricity came from renewable sources. The rest came from diesel installation run by another company, Llanos Blancos. The problem is that the supposedly reliable Atlantic trade winds are not so reliable or steady in the months of September through to February. Despite the steep price tag for the pumped hydro system, especially for a customer base of just 11,000 residents, analysts suspect that it is still too small for the micro system to take more than 60 per cent of its power from renewables.

To make matters worse for renewables activists the electricity system is only part of the island's energy demands as there are several thousand cars on the island, and the many tourists who come to the island are advised to hire a car rather than try to walk the steep, windy roads. In 2011 it was estimated that another €50 million or so would be required to convert to EVs, including the cost of installing the charging stations. Nothing seems to have come of this.

What about King Island in Australia's Bass Strait, well out in the supposedly reliable trade winds, which uses a mixture of wind, solar and diesel to produce electricity? The island is too flat for pumped hydro but the King Island Renewable Energy Integration Project includes a gigantic lead-acid battery capable of supplying the island's 2,000 or so residents for 45 minutes. According to information sheets issued by Hydro Tasmania, which runs the project, the project aims for renewables to supply an average of 65 per cent of the island's electricity and may well achieve that goal. The island also gained a wave generator during 2022 which, for a few million dollars more, has been reported as supplying about 3 per cent of the island's power.

(It should be noted that a major advantage for the island and other micro-grids is that they are small enough to be supplied by diesel generators, which can easily be powered up and down by computer control to suit variations in the output of the renewable installations. Another point to note is that that the various systems mentioned in this chapter may have cost 50 per cent more to build due to the remote location.)

For those interested in such systems the KIREIP has its own easily found web site. When I looked at this one fine Sunday morning in Melbourne the contribution of wind varied from a small negative amount (wind generators have electro-magnets which require electricity) up to 30 per cent of the island's power, and back to 16 per cent in perhaps half an hour. The operating system adjusted the diesel stations' output to suit these variations.

This obviously saves on expensive diesel which has to be shipped to the island but are the savings really worth the $18 million or so cost of the project? Independent cost-benefit analyses are hard to find but it does not seem likely.

There are renewable success stories for island grids, but hydro plays a major role in those, and the cost can still be considerable. Kodiak Island off the Southern coast of Alaska, America's second largest island, now sources 100 per cent of its electricity from renewables, but it was at 80 per cent before the renewable energy era, thanks to plentiful hydro power from the island's Terror Lake facility.

However, the island's council was still paying out $US7 million a year for diesel to supply the remaining 20 per cent of power to a grid servicing 15,000 residents and the island's thriving fishing industry. Fortunately, the hydro facility could be expanded by the addition of another turbine. The council also built a wind farm and added a battery capable of storing about 30 seconds of power from the wind farm – enough to keep the network supplied when the wind dies and the hydro power is being ramped up. About 14 per cent of the island's power is now from wind with the rest from hydro. The diesel generators have been left in place, ready for emergencies.

Hydropower works well with wind, incidentally, as it can be adjusted very quickly to suit changes in the output of wind farms.

The upgrades required for the clean network cost the Kodiak Island council about $US55 million, including a $US14 million grant and nearly $US40 million in special clean energy bonds with nearly zero interest. Despite all that assistance the payback period has still been calculated at about nine years.

One island community able to put sunshine rather than hydro to work in going 100 per cent renewable is that of Tokelau, a New Zealand protectorate of 1,500 people living on three remote atolls in the South Pacific. Rather than pay out $US800,000 a year for diesel but still only get electricity 12 to 18 hours a day, the Tokelauans got a $7 million grant from the New Zealand government to buy PV panels, batteries and coconut oil plants to make the oil capable of powering diesels as well as the island's many outboard motors.

The system was degraded by the harsh climate, but an upgrade ensured the island had 24 hour electricity. The investment is expected to pay for itself within nine years.

To return to Australia what about the opal mining town of Coober Pedy, where the climate is so hot most of its houses are underground? The town, which is far too remote to be connected to the major grids, has a hybrid diesel, wind and solar system. The resulting power is about 70 per cent renewable on average, with reports putting it as high as 80 per cent. This must be the record for a micro-grid in Australia, and all for a mere $18.4 million. The town's 1,600-1,700 residents are not expected to pay for the system, at least not through their power bills. Electricity is billed by averaging power bills in other parts of South Australia.

In the cases investigated, the communities involved were trucking in large quantities of diesel, and renewables helped cut the consumption of that fuel. However, even then, the payback periods are too long to attract private investment. As noted, systems in such remote locations cost far more to build, and there would be economies of scale for larger systems. But the results are still far from encouraging. Any shift to renewables worldwide will require considerable government intervention.

10

ROAD TO RUIN
– ELECTRIC VEHICLES

* Electric vehicles have no market advantage, apart from green cachet. Also, at the moment, they are far too expensive in Australia for mass adoption. Overseas experience shows that consumers have no inherent preference for electric vehicles over their petrol driven counterparts.

* Getting consumers to adopt EVs in any numbers will require massive amounts of taxpayer money – in either direct subsidies or tax foregone.

* A major result of large numbers of EVs on the road will be to strain a power grid already struggling to cope with the demands of the so called transition to renewable energy.

* Despite the obvious problem in adopting EVs, countries world-wide seem to be vying with one another in wasting money in this area. Australia should not go down the same path.

The push to make Hydrogen as a green version of LNG may take the top prize as the worst idea in our environment-crazy world, but government insistence on foisting electric cars on drivers who mostly don't seem to want them is not far behind.

Despite the total lack of logic of loading many thousands of additional electric units onto power grids that are likely to be gravely weakened by a range of other dubious green policies, governments worldwide seem to have gone mad over the adoption of these vehicles.

All sorts of incentives are being offered to those who buy EVs and disincentives for those sufficiently old-fashioned to stick to petrol cars. The American state of California intends to ban the sale of new petrol guzzling cars by 2035, while the UK intends to ban them from 2030. France is aiming for a ban by 2040 while Norway, arguably the poster child for the adoption of EVs, wants to ban the sale of new petrol cars by 2025. Car manufacturers have jumped on this electrified band wagon by declaring that they will stop making petrol driven cars by some date in the future.

Just how many of these bans will come into effect remains to be seen. With the exception of Norway, the ban deadlines have been set for long after the present governments and even the successor governments, will be out of office. None the less in the interim there will be expensive policies to encourage consumers to buy EVs.

For an indication of the horrific costs involved we need look no further than Norway, which managed to boost the proportion of EVs in its new car sales to an astonishing 84 per cent in January 2022.

To get that sales result the government has abolished or reduced a host of taxes and charges imposed on new cars, including the Value Added Tax levied on all retail items, as well as the frequent tolls charged for the tunnels and bridges. Those tolls are a feature of life in Norway and can be a significant part of the cost of living. Parking charges have also been greatly reduced or abolished for EVs.

In addition, the Norwegian government has installed 16,000 charging points, including one every 50 kilometres of highway throughout the country, and has a power grid strong enough to cope with the additional cars. The country has so much hydropower, thanks to its geography, that it has been able to close its few coal power stations without regret, power the EVs flooding onto its roads, and is still able to export power to other countries.

Cost estimates are hard to come by, but the Financial Times in the UK has calculated that the revenue forgone by the Norwegian government amounts to about half the cost of the EV. Various suggestions for reducing these costs to tax revenue by cutting some of the concessions have come to nothing. In any case, the history of concessions offered in countries around the world, notably Denmark and China, indicates that when the incentives are taken away consumers quickly lose interest in electric cars and sales collapse.

A typical example of this occurred in Germany when sales of EVs dropped by 83 per cent, thanks to the government cutting support payments of 6,000 Euros per car. Of course, EVs still accounted for 15 per cent of sales in January, which is high internationally and plug-in hybrids also did well. However, that is also because there are still major incentives. Consumers in this area have to be enticed.

To put these consumer enticements in an Australian context, let us suppose the state and Federal governments offered incentives amounting to $10,000 for each electric car. About one million cars are sold each year in Australia, when there is no pandemic, and to keep the maths simple let us say that half of those are EVs. That works out to $5 billion a year. Then there are the costs of building charging points in a country many times the size of Norway, and of building up the power grid, already struggling to cope with everyday demands on it – struggles compounded by reliable coal power plants falling off the grid and large amounts of intermittent power coming onto it. Yet the above policy would require half a million cars to be added to the demands of the grid each and every year.

That huge extra load which must be met somehow by installing yet more PV panels and wind farms although, as noted in earlier chapters, these are not being built in sufficient numbers even to de-carbonise the grid and gas turbines will still be needed. If the power used by an EV is even partly supplied by fossil fuels there would seem to be little point in using such cars. All the electric car is doing is shifting emissions from the city street to the power plant. On top of all that is the problem of installing charging points for cars usually parked in apartment building car parks. The change may require a significant upgrade to the buildings electrical connection at substantial cost. What about cars that are usually left out in the street overnight, as happens in some areas?

As can be seen from that basic scenario the bill for luring consumers away from their petrol-using cars, and then keeping those cars powered, quickly adds up to vast sums. In Australia, that is money not being spent on initiatives that might make a difference such as improved public transport, electric buses and a railway freight network to take more of the transport task from trucks. Also note that all the talk is about passenger vehicles, which is only a part of the fuel consumption story. Heavy vehicles can account for up to half of the fuel consumption of advanced countries but there is, as yet, no feasible way to electrify long-haul freight transport, either in trucks or by rail. Cargo ships and aircraft have also resisted electrification.

One major factor in restricting EV sales in Australia is the lack of supply, particularly of cheaper models. The best selling EV in Australia is far and away the Tesla-3, with a range of 500 kilometres retailing for $62,500 (prices are at the time of writing). The also popular Nissan Leaf sells for $53,000 drive away.

A few shipments of somewhat cheaper cars have been received and sales have reached 3 per cent of the total market, but car manufacturers still prefer to ship to other countries as car emissions policies elsewhere give retailers a major incentive to sell EVs. This is a long story but it involves car makers being required to meet certain emission levels, averaged across their total sales, or else. There are no such incentives in Australia. Cars on sale must meet certain emission standards but that is a different matter.

Assuming that the supply problem can be overcome and EVs become cheap enough to fit the budget of average consumers, what reason would average Australian consumers – as opposed to the rich and green conscious – have for buying EVs? Activists in this area point out that such vehicles cost less to refuel and less to maintain and repair over the vehicle's life cycle, glossing over the problem of re-sale (trade-in) value, the complication of replacing the battery after a certain number of years, and the constant need to ensure access to charging points. There is anecdotal evidence from Norway that many families are sufficiently worried about recharging to keep an electric car for commuting and another, aging conventional car for serious travelling.

As EVs have no market edge or advantage over petrol cars and a marked disadvantage in recharging, it is not surprising that consumers have to be given serious inducements to buy them. Then there will be the additional, vast expense of strengthening the grid to cope with large numbers of new power consuming units, and of building recharging points everywhere. All this adds up to a huge expenditure of taxpayers money for no real reduction in emissions.

In the meantime, as there is little supply available in Australia, most of the announced government initiatives in this area will do little but give money to those who would have bought EVs anyway – the rich, green-conscious consumer. EV policies are a straight waste of taxpayer money.

11

FOSSIL FUEL FURPHIES

* One of the more absurd aspects of the demonisation of fossil fuels which has contributed to the ongoing energy crisis is the assertion that Federal government subsidises fossil fuels.

* Nothing could be further from the truth. Mining and gas extraction is heavily taxed in Australia. The "subsidies" which activists point to are, in fact, either tax concessions available to all industries or refunds, again available to all industries. Instead of the government paying the dreaded fossil fuel producers money to keep going, as the term "subsidy" implies, they are just collecting less tax.

The assertion that fossil fuels are somehow being subsidised by the Australian government is a part of the on-going, often irrational campaign against the likes of coal and gas – a campaign that has badly affected our power grids. These claims include often wildly inflated figures for tax refunds which are paid to mining and gas companies and omit the salient point that those companies pay many times the refund in tax.

The lunacy of assertions that fossil fuel producers are subsidised reached new heights during the Federal election campaign in May 2022 when the sole Green party member of the Federal House of Representatives, Adam Bandt, claimed that the Federal government was handing $10 billion a year in subsidies to the fossil fuel industry. Even a fact checking body used by the broadcaster (RMIT ABC Fact Check) which is tolerant of various slips by the ABC, had to admit that the claim was "overblown".

However, it is possible to find vocal people at social gatherings, journalists and even academics who should know better also making this claim. So deep is the prejudice against mining and gas extraction that when confronted with the facts these activists will at best sullenly refuse to correct the claim, or at worst come up with all sorts of bizarre reasons as to why it's really correct.

The Minerals Council of Australia estimated that in 2019-20 the minerals industries (that is, minerals of all types but not LNG) in this country paid about $40 billion to state and federal governments. This included both company tax (a Federal tax) and mineral royalties (imposed by the states).

In Australia, all resources are owned by the governments, not by the individuals who own the land on which the resources are found. Broadly, the Federal government imposes royalties on LNG extraction while the state governments charge royalties for minerals. The Federal government also taxes company profits. For mining companies, the royalties are charged as a percentage of the market price for each tonne of mineral extracted, where that percentage can also change on a sliding scale depending on the price.

As prices for both coal and LNG have skyrocketed, thanks to activist campaigns against new projects, both company tax receipts and royalty payments have since increased markedly, helping state and Federal governments repair budget deficits. The Queensland state budget handed down in June 2022 opted to increase royalties sharply for the state's large coal mining industry, in spite of complaints by both the miners and the Japanese government, as Japan is a major customer for Queensland coal. It is estimated that the changes will result in an additional $8.5 billion revenue for the Queensland government. So much for claims that the industry is subsidised.

One point often brought up by those determined to find subsidies where they don't exist is that of the fuel tax rebates permitted in Australia. The government imposes an excise on both diesel and petrol to recoup some of cost of maintaining roads. Industries that use fuel to run vehicles and boats that do not use roads, including the heavy vehicles used by the mining industry, are entitled to claim back the excise. For various reasons it is administratively simpler for those who use off road vehicles to claim the excise back, rather than buy the fuel excise free in the first place.

The Federal budget papers for 2022 estimate that in 2021-22, the excise on petrol will amount to $6.1 billion with another $13.1 billion to be raised from the excise on diesel. About $8.5 billion is to be refunded, with perhaps 40 per cent of the that amount to be claimed by the mining industry. The rest will be returned to a range of industries including ferry operators, the fishing industry and agriculture (tractors and large vessels use diesel).

Taking a step back, subsidies have rarely been used as a form of industry assistance in Australia. Instead, governments have used indirect assistance such as import tariffs (manufacturing) or restrictive marketing arrangements including the likes of egg and meat marketing boards (agriculture), all of which involved consumers paying more. The protection system for agriculture, a major support base for the old Country (now National) Party became so extensive that it was referred to as Agri-socialism. These forms of protection were dismantled from the mid-1980s on, starting with the Hawke-Keating governments. Although there are vestiges of the system left, it is nothing like the old protectionist approach. Crucially, the minerals and gas and oil industries were never protected as they never faced the same import pressures as manufacturing or had the same voter base as agriculture.

The really concerning part about the subsidy nonsense is that it is not only obviously 180 degrees wrong but rarely corrected.

ADDENDA

Activists and even sensible people who should know better are convinced that all the opposition to renewable energy is really a sinister plot by big energy companies to keep the economy hooked on fossil fuels. It seems these companies are paying people such as this author to point out the obvious – that cheap, green energy is expensive and not very green. I only wish! Where is all this secret money and why can't I have some?

The reality is, of course, that this conspiracy does not exist mostly because there is no reason for it to exist. The local electricity market is only a tiny fraction of the overall market for big energy, as activists understand the term. The major coal-fired power stations are (or were) mostly brown coal which isn't exported. Brown coal mining is part of the electricity sector, not big energy. There are black coal power stations, but their consumption barely counts beside exports to the likes of Japan, China and Korea. In any case about 60 per cent of Australian coal production is metallurgical coal (for smelting steel), almost all of which is exported.

As a result, the major energy companies have rarely been sighted in the renewables debate mostly because it isn't their problem, and their only reward for becoming involved would be vicious attacks from the green industry. If they say anything it is to protest that they are really green. Mining sites usually have their own, tiny grids and these days those grids usually have a renewable energy component, for what that is worth.

For the 'show me the money' crowd the best option is to go green. As this book was being written one Melbourne council was advertising a position for a climate emergency engagement officer, who would earn more than $90,000 a year for working four days a week. This officer will report to the manager of a sustainability unit, no doubt staffed by well paid council servants. I am sure that all these officials will labor hard for the residents of their local government area, but this still adds up to serious money and a major support base for green policies.

What about the Non-Government Organisations? The published accounts for Greenpeace Australia Pacific Ltd alone, show total revenues of $22 million for 2021. For that matter what about the endless government funding for green projects of all sorts, and positions like climate fellowships at universities. There is nothing like that available to anyone who expresses scepticism about renewable energy – no wonder the debate has gone completely off the rails.

12

WASTING MONEY FOR A CAUSE

* All efforts at emissions reductions by Australia, including the mooted mass adoption of electric cars and forays into renewable energy are a straight waste of money.

* As major emitters such as China and India have shown no real interest in reducing emissions no matter what activists might say, Australia's potential tiny contribution will make no difference one way or another.

* Even if there was a concerted effort to reduce emissions by all nations, it is still difficult to construct a cost-benefit case for any action on emissions – where a dollar spent on emissions reduction earns back a dollar (that is, the equivalent of a present day dollar) in climate damage averted.

* In all cases a dollar spent on adaptation and disaster prevention, such as flood mitigation, burning off to reduce potential fire hazards in the bush, and improving housing stock to reduce damage from tropical storms, will earn more than a dollar in damage reduction due to natural hazards. The climate models are not relevant.

When the US House Speaker Nancy Pelosi (now retired)

visited Taiwan in August 2022, China expressed its displeasure by a number of measures, including announcing that it would suspend any co-operation with the US in the fight against climate change. This left commentators scratching their heads as there had been little co-operation to begin with. China generates more emissions than the rest of the industrialised world combined yet all it would commit to under the 2015 Paris agreement was for its emissions to peak in 2030. Since then, while making promises to curtail emissions, the country has been binge building coal fired power stations.

Activists have seized upon the fact that China is also building plenty of wind turbines and solar farms as "proof" that the country is serious about climate change pledges, but this is meaningless. China is still accounts for half the world's production of coal and more than half the world's production of steel and cement (both industries are a major source of emissions), with no indication that they are about to retreat in any of these industries.

In fact, as shown in the chapter on emissions, very few countries are bothering with meaningful reductions in emissions. But even if almost all countries involved in the Paris treaty can be persuaded to actually do something about emissions, as opposed to simply talking about the issue, it is still very hard to justify spending money on reducing emissions. Just analysing the effects of money spent on emissions reduction has always been a enormous problem.

For a start, the IPCC reports generate a very wide range of forecasts. Each report, including the most recent one of 2021, includes four emissions scenarios ranging from extreme to mild. Each of those scenarios in turn generates a range of possible outcomes.

Take sea levels for example. The IPCC 2021 report forecasts anything from an increase of 0.3 metres (about a foot in the old Imperial measures) by end of the century to one metre. Those keen to push the renewable energy cause will claim that everything is so bad that current increases are at the top of the projected range. That is certainly not the case for changes in sea levels, which have been tracked month by month by satellites using lasers since the early 1990s, with the resulting figures publically available on a site maintained by Columbia University in the US.

This shows the average sea level increase to be around 3.3 millimetres a year which, if continued for a whole century, adds up to about one third of a metre. In other words actual measured increases in sea levels are running at or below the bottom end of the forecast scenarios, not the top. The IPCC report makes all sorts of forecasts and claims about how the melting of the ice in Greenland or the Antarctica will accelerate the rate of sea level increase, but then so have all the other reports since the first one in 1990 with nothing much happening.

Readers who doubt this can always conduct their own investigation. They or their families may have a favourite beach, and will probably be able to find serviceable pictures of the same beach dating back to at least the 1950s, if not earlier. Does it look as if the beach has changed much? There are clear pictures of Fort Denison in the middle of Sydney Harbour taken 150 years ago. No change.

Much the same argument can be made about temperature changes. Although there has undoubtedly been some warming since 1990, when the first IPCC report was issued, this has proved far short of the forecasts in those earlier reports. In its annual State of the Climate report for 2021, the American Metereological Society, very much a pro-warming organisation, estimates the world to be warming at a rate of 0.18°–0.20°C per decade, or maybe one degree for every fifty years. (A basic Excel analysis by me of the global temperatures compiled by Haley Climate Research Centre in the UK since 1990 gave a similar result, for what that is worth. If the start date is set at 2000, the average increase is much lower. The increases have not been accelerating – just the opposite.)

Temperature forecasts issued by IPCC in its 2021 report, *Summary for Policymakers*, are expressed as an increase from an average of the 1850–1900 global surface temperature. The forecast is that the average global surface temperature for the years 2081–2100 will be higher than the 1850-1899 average by anywhere between 1.0-5.7°C, which is such a broad range as to be almost completely meaningless.

However, as it is widely acknowledged the world has already warmed about a degree since the 1850s (maybe somewhat earlier), that forecast is really saying that the temperature increase from now to the end of the century could be anything from nothing to 4.7 C. This is still completely meaningless, but at least current trends are not right at the bottom of the forecast range. (Please note that this book does not challenge global warming theory. It uses the figures issued by the IPCC.)

But if we take the existing trends at face value isn't an increase of a degree in fifty years or so a serious matter and aren't we in the midst of a climate crisis, or so we are constantly told?

One sure indication of any climate crisis should be the agriculture sector, shouldn't it? The Australian Bureau of Agricultural and Resource Economics and Sciences, a part of the Federal Department of Agricultural Fisheries and Forestry, says that there has been a general reduction in rainfall in Southern Australia in the past 20 years and a general increase in temperatures. Fair enough. However, the Agriculture Outlook produced quarterly by the Bureau points only to increases in production, despite the change in climate. The September 2022 quarter includes graphs showing that gross agricultural production has not quite doubled in a decade, although exports have doubled. As this book is being written this boom time for farmers is being badly spoiled by a steep rise in costs, notably the cost of fertiliser in part caused by the ongoing Russian-Ukrainian war.

A far more detailed analysis of production figures published in the same journal in 2019 shows that in real terms (that is, adjusting for inflation) farmers have boosted production by 50 per cent in 30 years. The severe droughts of 2003-04 and 2007-08 in the South East affected agriculture substantially at the time, but production bounced back and continued to climb, with no apparent permanent effect. Floods and fires do not seem to affect overall agricultural production

In other words, farmers use their considerable experience in dealing with droughts to ride them out and are continually adapting their techniques and farming practices to increase production. Of course they might be concerned over climate, but climate is just one of several factors that farmers deal with.

As for the fires and floods, sure there have been plenty of those but then Australia has always had floods and fires. For that matter since when are bushfires directly the result of changes in temperatures? A bad bushfire season will typically require a season or two of wet conditions in order to build up fuel in forests, and then a hot season or two to dry it out. Crucially, the residents of bushfire zones and their governments also have to avoid doing anything about burning off the excess fuel during the cooler months, to set up the conditions for a really bad bushfire season.

Determined doom sayers typically brush all this aside saying that its "obvious" that the climate is in crisis and that it will all get worse over the next twenty years or so. The trouble is for results that far in the future the time value of money overshadows all investment decisions. Broadly a dollar spent now is worth vastly more than a dollar in benefits gained in, say, twenty years' time. If the dollar is invested for a reasonable return of 5 per cent (ignoring inflation) after twenty years the nest egg is about $2.71. If we invest $1 trillion now – an optimistic estimate of the investment needed to get to net zero – then that investment has to avert $2.71 trillion worth of damage in current dollars in twenty years just to break even. Of course that also assumes all other countries will reduce emissions, instead of just talking about reducing emissions.

To make matters worse for climate activists the return on a dollar invested in reducing emissions should be compared with the return on a dollar spent on adaptation – say a dollar spent on developing strains of wheat that thrive in higher temperatures or in flood control works, or in developing better storm and flood warning systems.

Such investments will pay off in reducing the effect of the natural disasters that have always been a feature of Australian life, no matter what happens with climate. Crucially they also do not depend on what other countries chose to do about emissions. Australia will still get the benefit. They are a much better choice.

But as demonstrated by the treatment of Stuart Kirk head of responsible investment at the asset management arm of London-based financial services group HSBC over a presentation he made at an event, the Financial Times Moral Money Summit, in late May 2022, proper economic analysis of investment in renewable energy is not the issue.

Kirk's presentation, entitled "Why investors need not worry about climate risk," pointed out that most of the projections of economic loss due to climate change either have to fudge the figures or come up with numbers that are too small over the long periods involved to matter at all.

He said that the IPCC's worst case scenario calculated a loss of 5 per cent of world GDP by 2100, but without mentioning that average world economic growth is 2-3 per cent, so that the world economy will have grown by between 500 to 1,000 per cent by then (remember growth compounds over time). The loss from climate change is negligible.

To try and get numbers that might matter, Kirk said that central bank climate risk assessment studies usually throw in a policy shock such as a major increase in carbon tax or a huge interest rate increase. This policy shock was never made explicit but could be found if analysts went "right to the back" of the study document.

Among other valid points in his presentation, Kirk likened the climate crisis to the Y2K bug that predicted a widespread computer glitch at the turn of the millennium – predictions that turned out to be wrong – and declared that "unsubstantiated, shrill, partisan, self-serving, apocalyptic warnings are ALWAYS wrong".

Kirk's presentation was approved internally by HSBC but proved far too honest for those who had been pushing the climate crisis story, and the executive subsequently resigned saying that his position had become "untenable". None the less climate studies have long had to fudge the figures to get results that are sufficiently scary.

In 2006, distinguished economist Nicholas Stern (now Lord Stern) conducted a study for the British Government entitled *The Economics of Climate Change*. This review found that Global GDP would be permanently reduced by between 5 and 20 per cent by climate change, but the costs of reducing greenhouse gas emissions could be limited to 1 per cent of global GDP each year.

The review caused quite a stir at the time, but independent assessments showed that it made a number of conflicting assumptions about future conditions, all designed to make things worse, including assuming the maximum forecast for temperature increases at the time. However, the key choice was a discount rate (the after-inflation investment return mentioned above) which was set at just 1.4 per cent. That meant the value of a dollar invested now is not much different to the value of a dollar in damage averted decades from now. Commentators at the time pointed out that just increasing the rate to 2 per cent halved estimates of future damage, and a rate of 4 per cent destroyed the case for economic action entirely, particularly as Stern was looking at likely damages two centuries out.

The Stern report, now sixteen years old, called for immediate action, but then all those reports call for immediate action. One of many in that long list is a June 2021 report by the Swiss Re Institute (Swiss Re is an insurance group) warning that climate change could wipe up to 18 per cent of GDP off the worldwide economy by 2050, but only if global temperatures rise by 3.2°C. As noted in the brief discussion of temperatures above, this is unlikely. In any case 18 per cent works out to only about one quarter of total economic growth, and to achieve that the world economy has to be wrecked now so that there is no economic growth at all. This isn't going to happen.

Another approach is that of distinguished economist and winner of 2018 Nobel prize for economics, William Nordhaus, who has little time for the existing approach of nationally determined goals for emissions reduction. He wants global action to raise the price of CO_2 emissions – a carbon tax – as the only effective way to reduce them. He may be right, at least about a world needing a carbon tax to reduce emissions, but that isn't going to happen.

The IPCC, for its part, seems to have given up making any definitive statements on the issue of economic damage. A 2022 report by the panel says that "the wide range of global estimates, and the lack of comparability between methodologies, does not allow for identification of a robust range of estimates" of the impacts of climate change.

Activists try to get around the obvious point that there is no economic case for reducing emissions and that the money spent fighting climate change is being wasted, by saying that Australia cannot stand by and do nothing while climate changes. Perhaps, but that is a moral choice which the community must make, while being fully aware that the money spent on curtailing emissions is just money out the door for no benefit.

An extreme example of the distortion in decision making due to the fixation on emissions occurred in Western Germany in July 2021, when more than 160 people died and many more were injured by flooding that had been correctly forecast by scientists days before the event. The authorities in some areas were slow to react and there was no proper flood warning system in place. The German government had been too busy worrying about climate change to concern itself with petty details like flood warning systems.

No matter how the calculations are done, money spent on reducing emissions is, in effect, being wasted. Money spent on adapting to the effects of natural disasters will, at least, gain some benefits. Instead of acknowledging this obvious reality, media commentators and politicians have been pushing for emissions reduction efforts irrespective of the costs. But why should they care, their jobs remain secure.

13

END TIMES

* Forecasting the end of the world did not start with Extinction Rebellion. There is a long history of forecasts of the end of the world and collapses of various types.

* To date, none of the repeated forecasts of the end of anything have proved right.

* Activists who claim that the world is about to end because of a climate crisis are not basing their claims on anything other than hysteria, or perhaps statements by individual scientists. There is nothing in the official IPCC reports to suggest that the world will end.

One of the stranger results of the growing obsession with climate – a result fuelled by the now receding Covid pandemic – has been to breathe new life into the cult of end times, the urge to believe that the end of the world is just around the corner. This feeds into the debate on the economics of climate change, with activists screaming that we must wreck the economy right away to avert the end of the world.

Among other strange results of this climate alarmism school children, who have not heard such grim warnings before, have been reduced to tears over what they believe to be their lost future.

This belief that the world is about to end is an old one. American survivalists have been digging bunkers and buying guns in preparation for a major breakdown in society for decades. Fringe Christian groups also frequently warn their followers that the end is at hand, with the most notorious example being that of Californian radio show evangelist Harold Camping who repeatedly set dates for the Christian Judgement Day. He gained millions of dollars in donations, only for each date to come and go without anything happening. He died in 2013, with no Judgement Day in sight.

End Times is not exclusively about climate, and not necessarily about the end of the world. Up until the start of the fracking boom in 2008 or so, there were plenty of people convinced that the world was going to run out of oil.

Writing this brought to mind the time I interviewed the Canadian zoologist, geneticist and prominent environmental activist David Suzuki back in the mid-1980s (yes, I was a journalist even longer than that). I have forgotten why I was sent to interview Dr Suzuki, now in his 80s but a prominent activist at the time who happened to be visiting Australia, but I remember that during the interview he asserted that the world's oil would run out by the year 2000.

One of my jobs at the time was to compile a weekly list of oil wells being drilled for my newspaper, a thankless task, and I was intrigued enough to ask him what he based this assertion on.

"Every report we have says that we'll run out of oil by then," he said.

"Really, can you name a report so I can look at it?" I said.

"Every report says this."

"Then it should be possible to easily name one."

"Every report we have says this," Dr Suzuki said.

I gave up and moved the conversation on but later spent some time looking at the issue. In fact, no report with any credibility at the time said that. There were a few that said oil production would "peak" or "plateau" (meaning get to the peak and stay there) in 2000, but nothing about oil production ending abruptly. Somewhere along the line activists had substituted the word end for peak, as that suited their apocalyptic world view.

This belief persisted, however, and was given a new lease of life early this century when disruptions in the oil market caused the price to spike to above $US100 (the reasons for this are still unclear). Then the fracking boom hit, and America transformed from a major oil importer into a major exporter, almost over night. Despite activists becoming nearly hysterical in denying the new reality that there is plenty of oil, claims that oil production will end at some date in the future have now vanished from the public debate. Instead, there are claims that the adoption of electric cars will cut demand. But it is still possible to hear people at social gatherings claim that oil production will end.

In the 1960s there were fears about over population and mass starvation in the 1970s which did not happen, and a widespread fear that the world's computers would stop working in the year 2000. Nothing happened. Scientists have repeatedly warned that Australia's Great Barrier Reef is about to die off – warnings that pre-date the climate saga. The reef is still there for tourists to look at. Vanuatu and other pacific islands were supposed to have disappeared beneath the waves long ago. They are still visible. The arctic should have melted by now. It's still there. Climate tipping points are always five to ten years away, but never seem to happen.

In 2009, for example, Australia's then chief scientist Prof

Penny Sackett declared that the planet had just "five years to avoid disastrous global warming". Thirteen years later and, the occasional flood and fire aside, to non-activists the planet seems fine.

Despite this long history of bung predictions, climate doomsaying is as healthy as ever with otherwise quite sensible people convinced that the end of the world, or perhaps a climate disaster, is just a few years away. Failing that they claim the world is already in a climate emergency and that bumper harvests don't matter.

One near relation of the climate catastrophe idea is that of Collapsology (not an English word, yet) or *Collapsologie* in French, with the French proving particularly enthusiastic about these warnings of society's future. Like their climate brethren, the apostles of Collapsology are largely immune to reason or critiques of source material.

One of the foundation books for this doomsaying is Jared Diamond's *Collapse – How Societies Choose to Fail or Survive* published in 2005 which contains a series of essays on various societies that collapsed, including those of Easter Island (destroyed its own environment), the Norse in Greenland (failed to adapt to a changing environment) and material about Australia where Diamond, a professor of Geography, spent some years.

However, scholars are dismissive of much of Diamond's book. There were initial suggestions that the Greenland Norse (Vikings who colonised the country for centuries) died out because they failed to adapt as the Medieval Warming Period ended and the climate cooled. Scholars now say that Norse simply abandoned the colonies for a host of reasons, including a collapse in the lucrative trade in Walrus tusk ivy, for the comparatively easier living of Iceland and Europe, space having been made for them by the Black Death.

Diamond's dramatic theories about how the collapse of society of Easter Island was due to competition by clan chiefs chewing up all the island's forests, never found any favour with academics. The ruling, largely undisputed theory is that the island's stone head building culture was obliterated by contact with Western diseases, slave traders and introduced species.

Diamond was also concerned about Australia's ability to survive and feed itself but as he was writing seventeen years ago and at one point eccentrically declares that the early settlements had such trouble feeding themselves that they relied on "food subsidies" sent all the way from England until the 1840s, it is difficult to take the book's material on Australia seriously.

End times marches on, however. In 2020 two French authors Pablo Servigne and Raphaël Stevens published a best seller, *How everything can collapse: a brief manual of collapsology for present generations* which a New York Times review describes as "building off" the work of Diamond. Then there is the also highly influential David Wallace-Wells's *The Uninhabitable Earth*. The publishing industry is doing very well out of the end of the world as we know it.

In June 2020 the UK Daily Telegraph reported that these books combined with the Covid lockdowns of the time seemed to have touched something in the French psyche in particular, with requests for training courses and membership of End Times Facebook sites skyrocketing. One of the archpriests of this movement is Yves Cochet, a former French environment minister who has taken to life of rural self-sufficiency in Brittany with a horse and cart for transport. At the time he told the French newspaper Le Monde that the collapse was happening faster than originally thought. Last time I checked French society was still hanging together, somehow.

Although the end of the Covid lockdowns seems to have robbed these apocalyptic predictions of their urgency, every now and again someone will pop up to say that climate trends are worse than expected and that a climate tipping point is just around the corner and so on, and on. Those involved in the likes of Extinction Rebellion are convinced that the world is going to end, and even mainstream politicians are occasionally quoted as saying that they are supporting emissions reduction policies as they want to save the world.

Right!

For the record, and for what it is worth, the official, agreed statement on climate set out in the latest IPCC report issued in 2021 does not say that the world is about to end. It says that there will be this and that dire outcome but does not forecast a major break.

Here is a fair sampling from the report's summary for policymakers. Bear in mind that when the reports talks of a 1.5 degree or two degree increase, it means 0.5 or one degree from now. As noted earlier in this book it is widely accepted that the world has warmed one degree since the middle of the 19th century.

Global surface temperature will continue to increase until at least mid-century under all emissions scenarios considered. Global warming of 1.5°C and 2°C will be exceeded during the 21st century unless deep reductions in CO_2 and other greenhouse gas emissions occur in the coming decades.

With every additional increment of global warming, changes in extremes continue to become larger. For example, every additional 0.5°C of global warming causes clearly discernible increases in the intensity and frequency of hot extremes, including heatwaves (very likely), and heavy precipitation (high confidence), as well as agricultural and ecological droughts in some regions (high confidence). Discernible changes in intensity and frequency of meteorological droughts, with more regions showing increases than decreases, are seen in some regions for every additional 0.5°C of global warming (medium confidence). Increases in frequency and intensity of hydrological droughts become larger with increasing global warming in some regions (medium confidence). There will be an increasing occurrence of some extreme events unprecedented in the observational record with additional global warming, even at 1.5°C of global warming.

Projected percentage changes in frequency are larger for rarer events (high confidence).

Okay, if we take this at face value, then bad stuff is going to happen with increasing frequency, and that bad stuff will be more intense, but none of it sounds like the end of life as we know it or that we should be hiding in caves, praying to be spared. Instead, perhaps we should persuade people to build their homes outside of flood plains or build homes that might cope with the occasional flood (stilts?), ensure that potentially flammable material is cleared from around buildings and power lines during bushfire off seasons, take steps to prevent heat deaths during extreme temperatures and redouble efforts to ensure settlements in cyclone prone areas are properly secured against storms.

Or do we need to concern ourselves about the last precaution? A paper in the journal Nature Climate Change in June 2022 authored by 12 mainly Australian academics states that the frequency of tropical cyclones has been declining due to climate change (*Declining tropical cyclone frequency under global warming*, Savin S. Chaud of the Federation University in Ballarat and others). Chaud has since commented that there is evidence that storms are becoming more intense.

This paper was reported straight faced by the mainstream media, largely without any acknowledgement that it contradicted decades of green propaganda.

All that said, every now and then scientists will declare that individual populations of various creatures such as bees or polar bears are in danger of extinction. Let's start with bees. About 2006 or so there were plenty of stories talking about sharp declines in bee populations, caused by an effect called Colony Collapse Disorder. Honey bees in particular are important for agriculture, and the reported loss of 20 to 40 per cent of bee colonies – where the bees desert their hives – each winter certainly sounded alarming. However, beekeepers have always lived with an average loss of 14 per cent and are quite used to repopulating hives with packets of bees bought online.

There is a US group called the Bee Informed Partnership (Sigh!) which, among other activities, tracks bee hive collapses. At the time of writing losses remain similar but the bee keeping industry is still getting on. The disorder has been attributed to the increased use of pesticides, increasing urbanisation affecting habitat, diseases and, of course, climate change. Declines in wild bee populations led to the inevitable warnings about eco-system collapse in 2015 but, seven years later, the eco-system is still there. Perhaps something else will tip it over the edge?

What about polar bears? Why have scientists repeatedly declared that polar bears are on the verge of extinction yet residents in remote Canadian and Alaskan town report seeing more of them? A World Wildlife Fund information sheet notes that polar bear species came under pressure due to hunting which remained unregulated until the 1970s. Now there are claims that the undoubted decline of summer sea ice in the Arctic will cause populations to collapse (that word again), as sea ice is an integral part of the bear's hunting style. This is ice which covers the seas around the Arctic in winter and then recedes in summer. Data compiled by the National Snow and Ice Data Centre, a part of NASA, which tracks the seasonal change in the article waters, shows that sea ice reached a record summer low in 2017 but each year since then, more or less, the summer low has increased. Those forecasting the end of Polar bears had assumed that the Summer low point would continue to decline until there was no more sea ice in Summer.

Then there are frogs. There is some cause for concern about the loss of frog species attributed to a disease in some regions, where that disease is inevitably linked to climate change in media stories. But if amphibians are bothered by variations in climate someone should get around to telling cane toads which have been in plague proportions in Queensland and the American state of Florida for many years now.

The parade of grim statistics of varying worth continues, with the 2022 Living Planet Index compiled by the Zoological Society of London pointing to an average decline of 69 per cent in studied populations of mammals, birds, amphibians, reptiles and fish since the 1970s. However, if you scratch around online you will find critiques of the index, including the Nature article *Clustered versus catastrophic global vertebrate declines* (by Canadian scientists, published online in November 2020) which says that the index is constructed so that a few outliers, that is a few species in real trouble, can greatly influence the result. Once those outliers are excluded, the trend switches to an increase. This paper has been strongly contested but pointing to 3 per cent or so of species in real trouble makes more sense, especially given that the likes of cane toads, rabbits and kangaroos are not about to go extinct any time soon.

End Times cultists may hold prayer meetings in their bunkers, but the rest of us should just get on with our lives knowing that neither Western Civilisation nor the coal industry, will die any time soon.

14

CONCLUSIONS AND SOLUTIONS

There are times when the public debate goes entirely off the rails. For those old enough to remember, a prime example is the Y2K bug when everybody became convinced that the world's computers would go haywire at midnight on December 31, 1999. The older computers and software would be unable to handle the changeover to 2000, as they had only two digits for the year or some such.

Anyone who thought that the endless talk about this computer crisis was overblown was simply ignored in favour of "experts" prophesising doom, the more doom the better. In the end, come the date, nothing whatever happened. Even the venerable 286 PCs of the time, which definitely had not been designed with the change in date in mind, were completely unaffected.

But the key point to note is that the "experts" who talked so loudly and long about disaster were never held to account for making forecasts that turned out to be completely wrong. They just turned up later talking about something else. Similarly, the many experts who made forecasts for numbers of infections and deaths from Covid which turned out to be wildly wrong were rarely held to account. There is no mechanism in the public debate for individuals who make such forecasts to be later confronted with their own words and asked to explain the discrepancy between their forecasts and the actual result. The media needs doom, gloom and disaster for headlines and there is always someone to oblige.

Activists hate any comparison between the ongoing endless doom-saying over climate and the Y2K bug saga. They say any comparison is grossly unfair or, those who remember the event, play down the immense fuss over the change in dates that occurred before the year 2000.

"There may have been a few stories," I was told once in an online discussion, "but not that much publicity."

Right! The billions of dollars consumed on remaking computer systems before the date were not really spent. Motorists did not queue at service stations on New Year's Eve 1999 to fill petrol tanks in case there was a problem at midnight. Householders did not buy packages of long-lasting milk to store in cupboards in case the power supply stopped working. It was all in our collective imaginations.

However, the real problem is not so much that doomsayers keep on making predictions of mass famine, over-population, the Great Barrier Reef being wiped out, holes in the ozone layer destroying life, the end of oil production and so on and on. The problem is that others make hasty decisions based on these forecasts without considering who is making them or what the science actually says, the realities of emission control or any other factor at all.

As far as the Australian grids are concerned this means that state governments and activists are wiping out a large part of the reliable generating capacity of the Australia east coast grid and replacing it with not very much, when it can be shown that this change will make no difference to the crisis they are hoping to avert. As discussed in this book, few countries are paying any attention to the Paris treaty which, in any case, is barely worth mentioning as a means of controlling emissions world-wide.

The situation is made even more absurd by the fact that the science, as contained in the IPCC reports, gives a wide range of forecasts, none of which end in a climate apocalypse that might justify the extreme measures noted in this book (assuming, for a moment, that those measures made any difference at all).

Granted there are this and that group of scientists pointing to eminent collapse of this or that eco system, or this or that wildlife population is declining. But as also pointed out elsewhere in this book, these mini crises have been a feature of the public debate since the start of the modern era of climate forecasting. This dates from 1988 when Professor James Hanson addressed a US congressional meeting claiming that the industrial emissions were causing warming. That is 35 years' worth stories about climate crises – bees, frogs, polar bears and much else are meant to be nearly extinct but are still there. The Great Barrier Reef should have vanished long ago but, again, is still there.

Here we should say a few words about one the major voices in the climate debate, Will Steffen, a professor emeritus at Australian National University when he passed away in January 2023. Professor Steffen's work is often cited by those who claim that climate trends are at the top of the forecast range. Whenever I have asked people to justify such a statement they have pointed me to a graph of his. Although I have no wish to challenge the expertise of Professor Steffen or Professor Hanson or any of the other distinguished scientists who, over the years, have declared that the world is in crisis, Professor Steffen seems to have adopted his own methodology in making a case for climate trends being at the top of the forecast range. Different choices would have substantially changed his analysis.

In any case, this book is not about climate as such but about asking why the rush? Why does the existing power network have to be torn down in so short a time, when there won't be anything to replace it with?

One answer seems to be that cutting emissions has become a goal in itself with the science an excuse or a justification for this apparent hatred of human emissions. When asked why they are attacking power companies those involved usually say that are doing it for their children or for the planet, although the only actual result will be to inflict suffering on millions of consumers and their children, who will be forced to pay much more for electricity that will no longer be delivered 24/7. If emissions are cut the activists are happy and never mind the consequences.

Another factor has to be party-political. One way for political wannabees to differentiate themselves in the party room is to demand tougher action on climate change, meaning tougher action on emissions. This escalates, with other groups demanding even tougher action, which seems to result in radicalisation – the wannabees start believing their own rhetoric. Party politics might also be a factor in the more insane demands by the Greens. At the time of writing the Greens in Parliament were trying to get the Federal Government to agree not to approve any more fossil fuel projects, in return for support for some part of their climate policies. This move is more about the Greens being seen as anti coal and gas, rather than expecting any Australian government in its right mind to make such a ban.

Former Labor prime minister Paul Keating who had a nice turn of phrase, would refer to the extreme left of his party as 'Balmain basket weavers' and the 'maddies'. Other terms used include 'Volvo Socialists' and 'Chardonnay Socialists'. Perhaps we should extend the generic term 'basket weavers' to the fringe of the wealthy political elite of both parties who are determined to inflict real pain on the less well-off in the name of climate. An extreme example is the climate alarm movement's poster child Greta Thunberg who, in early 2023, released a book, entitled simply *The Climate Book*, which declared that to fix the climate crisis modern life as we know it has to end. Ms Thunberg is known to have become rich telling others that they have to become poor.

Back in Australia, the more realistic minded sections of the political elites seem unable or unwilling to stand up to these climate basket weavers, perhaps for fear of being seen to be soft on climate issues. Instead of provoking outrage by pointing to reality – that even if the climate story is right, the changes will make no difference so why the hurry – all players seem anxious to appease these basket-weavers.

This brings us to possible solutions to this looming, self-imposed disaster. Given the reluctance of the mainstream in all parties to bring the zealots to heel, a high level of green energy production would seem unavoidable. We will be stuck with a load of renewables. But the problems resulting from so many renewable generators can be countered by building more firm capacity, such as coal plants, gas plants and nuclear reactors.

For those who want to reduce emissions then nuclear reactors are the obvious choice, but it is difficult to see how a reactor can be built in Australia in anything like the time frame required. Even finding a spot for a low level nuclear waste repository has proved hard enough in this country; a whole nuclear reactor would be several bridges too far. Nuclear submarines, at least, have proved acceptable. Maybe we can return to the issue in a decade or so, particular after consumers have been left in the dark for long periods by renewables-dominated grids?

A coal plant is also a tough sell, given the present hatred of emissions. However, gas turbines still seem to be almost acceptable and are much better at adjusting output to suit the erratic variations in power from renewables than coal plants. As noted elsewhere in this book, the government has been urged to do just that. The one at Kurri Kurri in NSW's Hunter Valley cost about $600 million and generates 660 MW. This was built by the former Morrison government with critics pointing out that initial estimates that it will only operate about 2 per cent of the time make it an uneconomic proposition.

That's right. It is uneconomic because activists, specifically the Victorian Government, have blocked the development of a capacity market in this country where generators are paid to be ready to fill in gaps in supply. These exist in the US and UK and European countries have capacity payments to encourage firm capacity to be available when renewable energy stops working. If the plant is not built what happens during the 2 per cent of the time it is not switched on?

Given that a large part of Australia's generating capacity is to be shut down over the next 10-15 years, Kurri Kurri may eventually be operating far more often than just 2 per cent. Perhaps ten more generating stations similar to that of Kurri Kurri may just keep the grid going. Consumers will still be at the mercy of gas prices but at least they will not be left in the dark and cold.

Another possibility is a lot of diesel generator power stations such as one built by US company APR Energy in just under two months for the South Australian government as a reaction to the major blackout of 2016 when the interconnector with Victoria was blown down in a storm (something similar happened in late 2022). No separate price tag for the project has been revealed but the array of generators can deliver a useful fast start capacity of 275 MWs. This is a much larger output than can be expected from a small modular nuclear reactor, for much less cost, but can be built very quickly and without any of the endless objections, legal challenges and protests that would dog a nuclear reactor. Nearby residents are not going to become hysterical over a diesel plant, as they might over a nuclear or a coal plant.

(Very briefly, modular reactors these are similar in design to the power plants of nuclear submarines and somewhat smaller than the submarines for which they were originally designed. It just has to be put somewhere out of the way, protected by a fence to keep out the stupid, and high voltage lines connected to it. The vendors will then tell the users to instruct the facility's computer system on how much power is wanted, but otherwise not to mess with the device, and that they will return in five years to change the fuel rods.)

It is of course ridiculous that our network of supposedly cheap, green renewables generators, pumped hydro storage and batteries has to be backed up by diesel generators but that is the reality of having to deal with the basket weavers. Generators relying on the wind and sun cannot deliver full power 24/7 and, unless we want to spend extended periods in the dark, the grid will need serious fossil fuel back up and that is the end of the matter. High power prices are the unavoidable result of this green madness.

Unfortunately, our political system is now full of people who either have become blind to this reality or are unwilling to admit the truth for fear of being relentlessly, viciously attacked by green activists, having their twitter accounts suspended for spreading "misinformation" and pilloried on ABC talk shows.

We may simply have to endure high power prices and blackouts of varying lengths before Green activists and politicians finally accept that renewables may not be the answer to everything. In the meantime, perhaps we can send our higher power bills to the greens and activists and ask them to pay.

www.ingramcontent.com/pod-product-compliance
Lightning Source LLC
Chambersburg PA
CBHW022059190326
41519CB00041B/1099